SOTHIS PRESS

Sothis Press aims to completely reimagine the public domain book with an emphasis on universal access to captivatingly designed print editions. We believe that technology will enable every single book ever published to be perpetually accessible in beautiful print form. Sothis is beauty in ubiquity.

www.sothis.press
1 Waterhouse Square, London, ECIN 2ST

The Plane-Table and Its Use in Topographical Surveying

From the Papers of the United
States, Coast Survey

U. S. Coast and Geodetic Survey

SOTHIS PRESS

ISBN 978-1-0368-0229-5

The Plane-Table and Its Use in Topographical Surveying is in the public domain.

This edition is published by Sothis Press © 2023. The company is registered in England and Wales. The Company Number is 14900389.

ON PLANE-TABLE

SURVEY REPORT, FOR 1865.]

ON

THE PLANE-TABLE

AND

ITS USE IN TOPOGRAPHICAL SURVEYING.

By

Alexander Medina Garrison

[FROM THE COAST SURVEY REPORT FOR 1865.]

WASHINGTON:
GOVERNMENT PRINTING OFFICE.
1870.

TREATISE ON THE PLANE-TABLE AND ITS USE IN TOPOGRAPHICAL SURVEYING.

[INTRODUCTORY NOTE.—The plane-table is used in the Coast Survey as the principal instrument for mapping the topographical features of the country, and is universally recognized as the most efficient and accurate means for that purpose. Its application under various conditions, the methods of its use, and styles of topographical representation, have received a great development in the practice of the topographers of the Coast Survey, and special acknowledgment is due in this respect to the comprehensive views, practical tact, and elegant taste of Assistant H. L. Whiting, whose efforts have established the high standard of topographical maps recognized in the Coast Survey. In order to meet the frequently expressed want of a treatise on the plane-table and its use, which does not appear to be supplied by any existing book in our language, the following essay has been prepared for this report by Assistant A. M. Harrison, who acknowledges his indebtedness to many of his colleagues for contributions and aid in its preparation. The chapter on the three-point problem has been supplied by Assistant E. Hergesheimer.]

The following description of the plane-table, as now used in the United States Coast Survey, and directions for its use, are given after a long test of its qualities on that work. Being the instrument best adapted to topographical field-work, the very inadequate notices given of it in most American and English works make it desirable to furnish topographical surveyors with a practical manual of its use. This paper may seem in some cases somewhat amplified, but those more familiar with the instrument will overlook details intended for the benefit of beginners.

The invention of the plane-table is ascribed to Prætorius in 1537, but the first published description appears to be that of Leonhard Zubler, in 1625, who ascribes the "beginning" of the instrument to one Eberhart, a stonemason. From this time forward it has received successive improvements, chiefly from the Germans and French, until it has reached its present form, which is sufficiently perfect for the nicest accuracy required in an extended topographical survey.

TOPOGRAPHY is that branch of surveying by which any portion of the land surface of the earth is mapped in plan on a specified scale or proportion of nature. With the plane-table such a map is constructed on the ground by at once drawing upon the paper, which is spread upon the table, the angles subtended by different objects, and determining by intersections their relative positions, instead of reading off the angles on graduated instruments and afterward plotting the lines by means of a protractor, as is done in other methods of surveying. The practice with the plane-table has in this respect a great advantage in directness and precision. The measurement of distances and of vertical angles are used, in conjunction with the method of intersections, to obtain all the data for representing the horizontal and vertical features on the map, which is drawn in the field with pencil, the details being filled in according to established conventional signs.

The plane-table of the Coast Survey (see Sketch No. 30) is composed of a well-seasoned drawing-board, with beveled or rounded edges, about thirty inches in length, twenty-four in width, and three-quarters of an inch thick. It is commonly made of several pieces of white pine, tongued and grooved together, with the grain running in different directions to prevent warping. It is supported upon three strong brass arms, to which it is fastened by screws passing through them and entering the under side of the board, the three holes for the reception of the screws being guarded by brass bushings, and situated equidistant from each other and from the center of the table. By means of these screws the board can be removed at will. The arms rest upon the sloping upper face of a conical plate of brass, to which they are permanently fixed. Upon its lower edge or periphery this cone is fashioned into a horizontally projecting rim, the inferior face of which is as nearly as possible a perfect plane, and this in its turn rests upon a corresponding rim of a somewhat greater diameter projecting slightly beyond it. This second rim forms the upper and outer flange of a circular metal disk in the form of a very shallow cylinder. The inferior face or plane of the upper flange or rim has, at its contact with the superior face of the lower, a horizontal rotary movement about a common center, which is the center also of the instrument, and the two are held together by means of a solid conical axis of brass extending upward from the center of the inner face of the lower disk. A socket of similar shape fits exactly over this axis, projecting downward from the inner side of the apex of the conical or upper disk. The two plates are held together by means of a mill-headed screw capping the cone from the outside, and which can be loosened or removed at pleasure.

A tangent screw and clamp fastened to the edge of the upper rim permit, when loose, the revolution of the table about its center, and, when clamped to the lower limb, hold the table firm while the tangent screw gives a more delicate movement.

Three equidistant vertical projections of brass, grooved on the under side, and cast in one piece with the under face of the lower disk, extending from the periphery toward the center, rest upon the points of three large screws which come through a heavy wooden block below. This block, which is the top of the stand and is approximate in form to an equilateral triangle, is made of three pieces or horizontal layers, and is two and a quarter inches thick and very strong.

The three screws last mentioned have large milled heads, are quite stout, and play through the block from below by means of brass female screws let into it. They are the leveling screws of the instrument, and are equidistant from its center.

Upon the under side and center of the lower metal disk is a socket containing a ball with a brass arm, which projects through the center of the block from beneath. The lower end of the arm is threaded, and upon it plays a female screw with a large milled head, which can be relaxed or tightened at pleasure. This screw clamps the whole upper part of the instrument to the stand; it is loosened only before leveling, and kept securely clamped at all other times.

The block is supported upon three legs, and with them forms the tripod or stand of the instrument, the legs being of such a length as to bring the table to a convenient height for working, and so arranged as to be taken off at will, or closed so that their iron-shod and pointed ends can be brought together or moved outward, as may be required. For lightness the legs are generally made open through the middle of their length, though sometimes they are solid, and each one is fashioned at the top into a cylindrical form with an outer flange, the cylinder fitting into a groove on the under side and near the edge of a truncated vertex of the block. The flange, by coming in contact with the lower edge of the block, prevents a too great spread of the legs. A brass screw, connected at right angles with the middle of a movable bolt which runs through the axis of the cylindrical head of the leg and projects through a hole in the block, is fastened above by a female screw with a large milled head.

A pair of compass sights or a watch telescope has sometimes been attached to the under side of the board of the plane-table. When the table has been put "in position," the watch telescope is directed to some well-defined object, and by after reference to it any movement which may have taken place out of position in the table during its use can be detected and adjusted. This, however, is but a complication of the instrument, and the same purpose can be more readily served by the alidade itself. The watch telescope has not been used in Coast Survey work.

Rollers have been attached to the under side of the table, taking the place of clamps for holding the map in its place; but these are very liable to get out of order, and are not regarded with favor by the best topographers.

The alidade consists of a brass rule about twenty-two inches long, having a circular level on its upper face. Near the middle of the rule is a perpendicular cylindrical column of brass, called the "standard," surmounted by two square brass plates joined by screws, and supporting, horizontally, a conical journal, through which extends a closely-fitting cone of brass, coming from and attached to the side of the telescope. This cone forms the axis of the vertical movement of the telescope, and is secured at the extremity by a screw which holds it in its place. The telescope itself has the usual cross-hairs and means of focal adjustment.

A transverse level is fastened to the edge of the upper of the two plates at the top of the standard by means of adjusting screws.

The telescope is so placed that its line of collimation is above and in the same vertical plane with the fiducial edge of the rule, though this is not absolutely necessary. Its position with regard to the edge should, however, be constant.

A vertical arc, with a tangent screw and clamp, is attached to the telescopic side of the lower brass plate, and, with a vernier which moves in arc as the telescope is raised or depressed, is used in the measurement of vertical angles for heights.

A small strip of brass is sometimes attached by means of horizontal hinges to the edge of the rule, after the manner of the ordinary parallel rule, for the purpose of obviating the necessity of watching the exact contact of the edge of the rule with the point while sighting; but as it requires accurate hinging, which is subject to wear, it has not come into general use. For reconnoissance for military purposes it is valuable.

A *declinatoire*, consisting of a rectangular metal box containing a magnetic needle and graduated arcs, the north and south line of which is parallel to the outer straight edge of the box; a scale of equal parts of brass or German silver; a set of metal clamps for fastening the map to the table; a pair of sharp dividers; India-rubber, pencils, and a pen-knife, complete the list of essentials for prosecuting plane-table work.

Since writing the foregoing description of the alidade, it has been somewhat modified and improved. (See frontispiece.) The telescope, instead of being supported on the side of the standard, is "transit mounted," resting upon an axis which is supported at either end upon a cross-piece of brass, upheld by two square columns, having their bases upon a square plate forming the top of the standard, the support and standard being firmly united. This arrangement brings the center of gravity over and coincident with that of the standard, instead of outside of it, as heretofore.

The vertical arc is better protected by being placed inside of the two square uprights, and the tangent screw made more stable by being made to play through one of the uprights on the opposite side from the arc.

An arrangement for attaching a riding level to the telescope, and the declinatoire being permanently fastened to the rule, are also improvements upon the old pattern.

ADJUSTMENTS.—From the nature of the service in some sections of the country the plane-table is often necessarily subjected to rough usage, and there is a constant liability to a disturbance of the adjustments; still, in careful hands, a well-made instrument may be used under very unfavorable conditions for a long time without being perceptibly affected. One should not fail, however, to make occasional examinations, and while at work, if any difficulty be encountered which cannot otherwise be accounted for, it should lead directly to a scrutiny of the adjustments.

1. *The fiducial edge of the rule.*—This should be a true, straight edge. Place the rule upon a smooth surface and draw a line along the edge, marking also the lines at the ends of the rule. Reverse the rule, and place the opposite ends upon the marked points, and again draw the line. If the two lines coincide, no adjustment is necessary; if not, the edge must be made true.

There is one deviation from a straight line, which, by a very rare possibility, the edge of the ruler might assume, and yet not be shown by the above test; it is when a part is convex, and a part similarly situated at the other end concave, in exactly the same degree and proportion. In this case, on reversal, a line drawn along the edge of the rule would be coincident with the other, though not a true right line; this can be tested by an exact straight edge.

2. *The level attached to the rule.*—Place the instrument in the middle of the table and bring the bubble to the center by means of the leveling screws of the table; draw lines along the edge and ends of the rule upon the board to show its exact position, then reverse 180°. If the bubble remain central, it is in adjustment; if not, correct it one-half by means of the leveling screws of the table, and the other half by the adjusting screws attached to the level. This should be repeated until the bubble keeps its central position, whichever way the rule may be placed upon the table. This presupposes the plane of the board to be true. If two levels are on the rule, they are examined and adjusted in a like manner.

Great care should be exercised in manipulation, lest the table be disturbed.

3. *Parallax.*—Move the eye-glass until the cross-hairs are perfectly distinct, and then direct the telescope to some distant well-defined object. If the contact remain perfect when the position of the eye is changed in any way, there is no parallax; but if it does not, then the focus of the object glass must be changed until there is no displacement of the contact. When this is the case, the cross-hairs are in the common focus of the object and eye-glasses. It may occur that the true focus of the cross-hairs is not obtained at first, in which case a readjustment is necessary, in order to see both them and the object with equal distinctness and without parallax.

4. *To make the line of collimation perpendicular to the axis of revolution of the telescope, and the axis of revolution parallel to the plane of the rule.*—The instrument is set up and carefully leveled, and the cross-hairs directed to a plumb or other vertical line. If the cross-hairs cover the line when the telescope is elevated and depressed, the adjustments are perfect; should they deviate, however, from the vertical line, this error may be attributable to two causes: 1st, the line of collimation is not perpendicular to the horizontal axis; or, 2d, the axis is not horizontal, and consequently not parallel to the plane of the rule. In the first case the motion of the cross-hairs will be in a curve, and upon being made to cover the vertical line when the telescope is horizontal, will deviate from it to the *same* side both upon elevation and depression. In the second case the movement of the cross-hairs will be in a straight line oblique to the horizon, and, when made to cover the vertical line when the telescope is horizontal, they will, upon being elevated and depressed, appear upon *different* sides of the vertical line. These two cases will be considered separately.

When the construction of the telescope admits of it, the perpendicularity of the line of collimation to the axis may be examined as follows: Direct the cross-hairs to a well-defined, distant object, as nearly upon a level with the telescope as may be, draw a line along the fiducial edge; then reverse the rule 180°, again placing the edge along this line, revolve the telescope upon its axis and again observe the object; if the cross-hairs cover it, the adjustment is perfect; if not, one-half the error must be corrected by moving the cross-hairs by means of the adjusting screws of the diaphragm, and the other half with the tangent screw of the table, and the operation should be repeated until the adjustment is complete.

In using the method just given, it may be taken for granted that the line of collimation revolves in the vertical plane of the fiducial edge, as any error arising from this not being the case would be inappreciable.

After this adjustment the horizontality of the axis should be examined. Direct the cross-hairs to a distant, well-defined, elevated or depressed object, having the table carefully leveled; draw a line along the fiducial edge, reverse the rule, and again direct toward the object; if the cross-hairs cover it, the axis is horizontal; if they do not, one-half of the deviation should be corrected by means of the screws attaching the upper plate to the top of the standard, or by means of the screws attaching the standard to the rule. The level attached to the axis should then be made central. In the alidades, as recently improved, the bearings of the axis being unchangeable, save by such violence as would destroy the instrument for all practical purposes, the foregoing adjustment and the succeeding one are, of course, unnecessary, as the instrument is to be considered as in constant adjustment in these respects.

5. *To make the line of collimation parallel to the vertical plane of the fiducial edge.*—The exact parallelism of these is not necessary, but it is essential that the deviation should remain constant. This adjustment may be examined by means of two needles stuck in the table. The table is turned so that the needles sight exactly to some distant object; the fiducial edge is then placed against them and the telescope directed to the object. If the cross-hairs bisect it, the adjustment is correct; but if they do not, it can be corrected by means of the screws attaching the standard to the rule.

6. *Zero of the vertical arc.*—When the line of sight is horizontal, the vernier of the vertical arc should read 0°, or the index error should be known. This may be examined by means of the distant sea horizon, or by setting up the alidade so that the center of the telescope is in the line of sight of an accurately adjusted leveling instrument, and then directing both instruments, while level, to a distant object; if any error be discovered, it may be corrected by setting the vernier at 0°, and adjusting the horizontal wire to the sea horizon or object.

When the above means are not available, the following method may be used: Set up the instrument at a point, measure the angle of elevation or depression of a distant object, remove the instrument to that object, and measure the angle of depression or elevation of the first point. These angles should be equal if the adjustment be correct; and if not equal, the index error will be one-half the difference of the two readings.

The following method of making this adjustment, where you have neither a separate level, a sea horizon, nor an elevation, may be employed: Set up the table and level it carefully on any level piece of ground between two equidistant points A and B, say 600 or 800 meters apart. Determine with the table the difference of level of these two points, and remove the table to A. Measure carefully the distance from the ground to the center of the axis of the telescope, and add or subtract this from the difference of level of the point B, according as it is lower or higher than A. Set up a target or distinct point at this height at B, direct the cross-hairs upon it, and correct the vernier accordingly.

A longitudinal riding level placed upon the telescope, or a level permanently fastened upon the top of the telescope parallel to the optical axis, and adjusted to the horizontal wire, will give the error at once.

Plane-table.—With regard to the plane-table proper, a disturbance of its good working condition generally arises from accidents resulting from carelessness or from undue exposure of the board to the inclemency of the weather, and where these injuries are of a serious nature the mechanician only can apply the proper remedies. A coating of shellac has been suggested, whereby the shrinkage and warping of the board is said to be prevented "in a very marked degree;" but well-seasoned wood and fidelity in construction must be the main reliance of the surveyor.

PAPER.—In addition to the faulty adjustment of his instrument the topographer has an additional source of error to guard against, arising from the expansion and contraction of the paper, due to its hygrometric nature. From the exposure to which a sheet is subject while in use in the field, and the occurrence of almost unceasing atmospheric changes, it can hardly ever be considered for any great length of time as fixed in its relative proportions; and the difficulty is greatly increased from the want of uniformity in this variation in the different parts of the sheets and in different directions.

In case of trouble with the points arising from this cause, there is but one remedy, and that is by the system of compensation as treated of in the article on field-work.

When points are determined by intersection, the effect of contraction and expansion may be uniform enough to be comparatively unimportant; but in running long traverses without side checks it is always felt.

In plotting long measured distances, the most feasible method of correction is to measure a minute of latitude near the place of plotting; and as the lengths of all these minutes on the sheet is the same, a comparison with the scale can at once be made and the percentage of error determined. When the sheet has no projection, squares constructed to scale upon it will answer the same purpose.

SCALES.—The very simple and ingenious decimal system of scales for maps adopted by the French is that in use by Coast Survey. In this system the scale of any map is represented by a fraction whose numerator is unity and whose denominator is some multiple of two or five, as $\frac{1}{20000}$, $\frac{1}{10000}$, $\frac{1}{5000}$, $\frac{1}{2000}$, meaning that any distance on the map is one twenty-thousandth, one ten-thousandth, &c., of its actual dimensions on the ground. Thus, on a scale of $\frac{1}{10000}$, one decimeter on the map will represent an actual distance of 1,000 meters.

Any other desirable scale can, of course, be used, as a given number of inches to a mile; and in case of triangulating from a base, as in a reconnoissance, no scale, even, need be adopted. By assuming two points on the sheet as the extremities of the base, and working from them, a correct delineation of the country can be obtained before the base has been measured. After measurement the scale of the map can be ascertained by dividing the length of the base on the map by its length on the ground, both expressed in the same unit.

In those regions where there is much detail, $\frac{1}{10000}$ is the scale generally used for field-work, while in others, where there is but little minute work, $\frac{1}{20000}$ is employed. Less than the latter is never used for field sheets. In some cases, such as surveys of cities, wharves, &c., $\frac{1}{5000}$, or larger, may be used; and in surveys for the location of batteries, the mapping of forts, navy yards, sites for light-houses, &c., scales as large as $\frac{1}{1200}$ have been used to advantage.

The diagonal scales of equal parts used on the Coast Survey with the plane-table, for the purpose of plotting measured distances, correspond with the scales of the maps. They are of metal, and sufficiently hard to stand long wear from the points of the dividers.

PROJECTIONS FOR FIELD-WORK.—The conical projection is that used in the Coast Survey for field-work.

The orienting of the sheet is determined by various considerations. It should include as many triangulation points as possible; it should duly conform to the position of sheets already surveyed in the same neighborhood; and it should embrace the area of the proposed survey in the manner most convenient for work, and most effective for the artistic appearance of the sheet when finished.

A sketch giving the triangulation points and the approximate shore line comprised in the area to be surveyed, being before the draughtsman, he proceeds as follows:

The limits of the sheet having been determined, the middle meridian A (see appended sketch) is located and drawn, and its intersection with the most central parallel determined, at which point the perpendicular B is erected.

The number of minutes of latitude on the central meridian, above and below the central parallel, being known, take the corresponding distance from Table VI, "Projection Tables," C. S. Report, 1853, Appendix No. 39, from under the head "Meridional Arcs," and lay it off (C) above and below the central parallel; and with the same distance as radius, strike arcs D D D D above and below from near the extremities of the perpendicular B. With a well-tested straight edge draw lines E E through the north and south minutes on the central meridian, and tangent to the two arcs D D, to the right and left. This gives three parallel lines perpendicular to the central meridian.

From the same Table VI, from under the head "Lengths of Arcs of the Parallels," take out the value corresponding to the number of minutes of longitude, east and west of the central meridian, and lay off the whole distance F F′ F″ on each perpendicular, taking each distance from its appropriate latitude. Subdivide these into minutes G G′ G″.

For the areas usually covered by plane-table sheets the corrections X, for determining the abscissas from the arcs of parallels, (Table VI, head "Co-ordinates of Curvature,") are inappreciable, and may be disregarded; the ordinates Y only being used. These give the distances to be set off from the lines B and E, perpendicularly toward the pole, for each minute of longitude counting from the central meridian. For ordinary field projections of scale $\frac{1}{10000}$ the ordinate of the extreme minute only need be used, and the parallel drawn a right line from the point so found to the central meridian. This ordinate H being set off on each of the parallels, the meridians are all drawn in with a fine ruling pen, then subdivided into minutes, and the parallels carefully ruled in through the points of subdivision.

The projection is verified by applying the measure of a number of minutes of latitude and longitude, and by a comparison of diagonal measurements on different parts of the sheet.

All measurements should be carefully taken from the scale with a keenly pointed beam-compass, and the marks pricked in the paper should be as light as possible to be seen, so as to insure the greatest possible accuracy.

The draughtsman is supplied with a list of triangulation points, which gives their relative distances, their latitudes and longitudes, and also the equivalents in meters of the *seconds* of latitude and longitude, according to which the points are now plotted on the sheet by measuring from the corresponding minutes. Thus in the diagram the distance J represents the seconds of latitude; K, the seconds of longitude of the trigonometrical point.

The accuracy of the plotting is tested by a measurement of the respective distances between the points with a beam-compass, these distances being also given. The latitude and longitude are then plainly marked, usually on the north and east sides of the sheet, at one extremity of each parallel and meridian, and the pencil marks erased.

It sometimes becomes necessary to base topographical work upon a detached scheme of triangulation, before the usual astronomical observations have been made. In this case the only elements given are the distances from the points to two projected arcs of rectangular co-ordinates, (which are asumed,) and the distances between the points. The projection for plotting these consists simply of axes of X and Y, so laid on the sheet that it will embrace all the points required by the surveyor, and in the manner most convenient for his work; and the points are plotted from these by the intersection of two arcs with the distances of the points from the axes as radii, either north or south, east or west, of the lines of X and Y, as the plus or minus signs given may indicate. The only test is by the distances between the points, and there should be at least two from each. If the work be correctly done, a conical projection can be constructed on the sheet after it is finished and the required astronomical work is completed.

Should it become necessary to make a topographical survey, when neither the data for projections nor co-ordinates nor table of sines and co-sines are at hand, plotting by distances is the only recourse left, and great care is absolutely necessary.

It has sometimes been found expedient to carry on a plane-table survey in advance of the triangulation, or where the triangulation has not yet been connected with a base. Under such circumstances it is advisable to draw squares of any specified number of meters on the sheet, by means of which the projection can ultimately be laid down correctly.

FIELD-WORK.—*General remarks.*—In organizing a party for field-work it is necessary to have one man to carry the table. His duty is to remain constantly with the instrument, to leave it under no circumstances; and while the topographer is at work he holds the shade to protect the chart from the glare of the sun. In some sections the labor of carrying the table is quite fatiguing, in which case another man should be employed with the shade. He should also keep the pencils sharpened, and sometimes, when a careful person, he levels the table, thus giving the topographer an opportunity to glance over the surrounding country. He should always have with him a spare piece of rubber, and one or two extra pencils. Two chainmen are needed, and two or three other men with signals, hatchet, telemeter, and other working apparatus to execute various offices, as they may be required. The maximum number necessary for field-work in a plane-table party on land is five hands, and when using a boat, six. Satisfactory work has been done, however, with three, and on very rare occasions with even two men; but, of course, with less facility. More than five and an aid, when but one table is used, is unnecessary, and a less number is a detriment to rapid execution.

The alidade is carried from station to station by the chief of party, resting on the bend of his arm, or hanging easily at the side, and in handling is to be seized by the lower part of the standard, never by the telescope or rule. Some topographers prefer to have it transported in the box by one of the men, and handed to them when the table is set up at a station. It usually weighs 8½ pounds; and there is a fear of its being put out of adjustment or injured by falls on rough ground, or in crossing insecure fences, if carried by hand, and a relief is afforded to the arm by being freed for a while from its weight; but one soon becomes inured to the weight so as to feel but little inconvenience from it, and carelessness in taking out and replacing it in the box so many times during the day is quite as likely to disturb its adjustments, as is also the fall of the box, or rudely setting it on the ground. The meter scale is best fastened under the clamps which hold the paper to the table, where it is close at hand ready for reference. It has been suggested that it would be an advantage to have it engraved upon the rule of the alidade, and it has also been proposed to have the scale drawn upon the sheet, and thus afford a correction of error of shrinkage, but its constant use would soon seriously deface the paper. The pencils, dividers, and India-rubber can be carried in an outside breast-pocket, the points of the dividers, when not in use, being thrust into the rubber. A little metallic pencil-holder, pinned upon the left breast of the coat, is used by some surveyors for this purpose. A handy and compact arrangement for carrying the scale, pencils, &c., is a russet leather case 10½ by 2¾ inches. It is made large enough to accommodate, on the opposite sides of the scale, when it is in, three or four pencils, and the dividers protecting the points of both; the whole carried in a leather pouch 11 by 4½ inches, slung over the shoulder, the pouch accommodating also note-books for sketching, table of heights, extra pencils, and rubber; everything being at hand and well protected. When the table is set up, the dividers and pencils are taken from the case and laid upon the table, and the scale drawn out as needed. Some topographers object to carrying the scale upon the table under the clamps, because it is liable to soil the paper, to drop out in passing from station to station, is not always in the most convenient place for use, and sometimes interferes with the play of the alidade.

It is well to have ready a light India-rubber cloth cover to slip over the board in case of a sudden shower, as well as to protect the paper from the dust on the roads, mud in swampy ground, or water where a boat is used in going from one station to another. The sides of the sheet, where they are turned under the table and come more or less in contact with the coat of the observer, should be protected by strips of paper about four inches wide, and six inches longer than the side of table, so as to fold under it and clamp on with the sheet itself. A plan followed by some topographers is to cover the whole sheet as exposed on the table with thin paper, tearing it away at those points only where they are at work, and covering again by pasting on patches as soon as finished, thus protecting as much of the sheet as possible; but in determining points by forward intersections this is impracticable.

The plane-table must never be rudely handled, never roughly set on the ground, nor carried heedlessly through woods or swamps; and the weight of the body or arms should never rest upon it. Instructions should be given the men that, under no circumstances, except in cases of threatened danger, should the table or instruments connected with it be touched during the temporary absence of the topographer.

Preliminary work.—As an indispensable preliminary to the operations of field-work, the topographer must assure himself of the correctness of the plotted points on the sheet, by an examination of them in the field, either by actual occupation of each one, or of a sufficient number to embrace them all in two or more lines of observation. When this has been done, and the points found correct, or properly adjusted, in the manner hereafter described, the regular survey is commenced.

When the determined points are too widely separated to supply, for all portions of the area to be mapped by the topographer, a sufficient number for good determination of position, it becomes necessary to determine others with special care by the plane-table. This is generally best done as the work progresses, and as the topographer develops his wants for points in the execution of details.

Sometimes, from lack of natural objects, it is found advisable to go beforehand over the country and locate signals in suitable points for subsequent determination and use. In the location of signals, either as permanent points or simply for temporary forward lines, a great deal depends upon the good judgment of the person placing them. Two purposes are to be subserved: first, the seeing of sufficient known points to give a good determination; and, second, to command a view of as great an area of country, and as many natural and artificial features for filling in the topography as possible. It should be remarked, also, that in the course of prosecution of the regular work, no favorable opportunity must be allowed to escape for locating a signal or determining a point which may at some future time be of service. Advantage should be taken of open places in the woods exposing roads or ravines. Piers or draws of bridges, or piles, giving lines up and down streams, with precipitous or bluffy and woody banks; trees of unusual appearance in prominent positions, or bearing flags placed upon them for the purpose; points of rock, off-shore or otherwise; lightning-rods, cupolas, weather-cocks, chimneys of factories, and other peculiar and marked objects come within this category. In fact, it may be set down as a rule, that well-determined signals located at convenient distances over the sheet are more likely to be too few than too frequent.

Signal poles should be straight and perpendicular, the flags upon them adapted in color to the background against which they will be seen when observed upon, and be protected from cattle in settled districts by stones piled or earth thrown up around their bases. They should also be well marked with pegs, or by measurements to neighboring permanent objects, so that in case they are disturbed their positions may be found.

It is taken for granted that some facility in the manipulation of the table is already arrived at, as well as a knowledge of conventional topographical signs (see Sketch No. 32) and the application of them. On maps of a large scale, it is required to plumb the plotted point exactly over the station, although on the usual field scale an approximation with the eye is all that is requisite. All lines should be drawn lightly and carefully close to the edge of the rule with a finely sharpened hard pencil; but in sketching one somewhat softer may be used. If the table and alidade be in proper condition, the contact of the fiducial edge with the paper will be perfect throughout its whole length; and in drawing a line along the edge care must be taken to preserve the same inclination of the pencil, and to avoid a "shoulder" in the pencil itself. If the rule be at all raised from the paper at any part, still greater care is to be observed lest the point of the pencil should run under the edge and thus deviate from a straight line.

It would be well for the beginner to learn to use his left eye as well as the right in sighting with the alidade, for obvious reasons.

The instruments should be kept scrupulously clean and free from sand or grit, and work with the table should cease the moment the presence of any foreign substance between the surfaces which play upon each other is suspected. An occasional taking apart of the table and cleaning with soap and water, using soft linen rags for the pupose, will be found necessary; and after being oiled and put together, it should be wiped thoroughly dry. The cleaning should not be intrusted to any person unaccustomed to the handling of instruments.

In observing upon signals which are not perpendicular, the sighting should be as nearly as possible upon the base of the pole.

Field practice.—Topographical points can be determined by three methods, viz: "intersection,"* "resection," and measured distances. In the first of these the point must be seen from two or more occupied points in suitable positions with regard to the point to be determined; in the second it must be occupied; and in the third there must be a direct measured line, with an established direction from the occupied point. These methods of determination, and the incidental operations which accompany them, will now be considered.

---

* Custom has given to this general term a specific signification in Coast Survey topographical work.

Fig. 1.

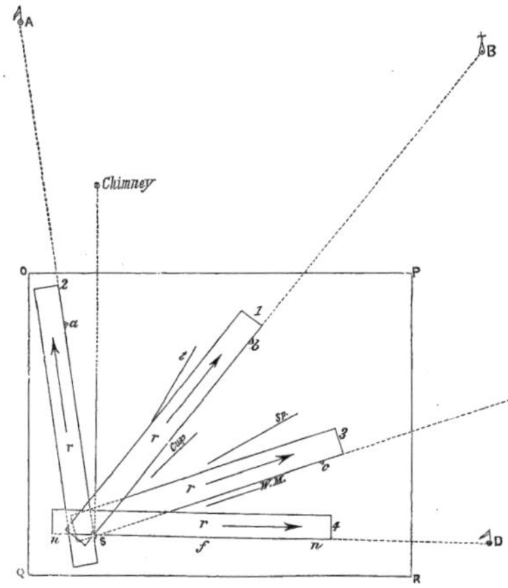

Let O, P, Q, R, Fig. 1, represent the board of the plane-table, upon which is spread the topographical sheet; the plotted triangulation point *a* upon the sheet representing the signal A upon the ground; *b*, the spire B; *c*, the signal C; and *s*, the station S; the small letters on the sheet representing the centers of the signals on the ground, which are referred to by corresponding large letters.

The table is first placed approximately level over the occupied station S, and put in position, also approximately, by the eye, so that the plotted points on the sheet are in range with the station S and the signals or objects they represent in the field. Then plumb the point *s* over the station S, fixing the legs of the table firmly in the ground; place the alidade upon the table so that the rule shall extend across its center; loosen the large milled head screw projecting below the top of the stand, and by means of the leveling screws bring the bubble of the circular level on the rule to the center. Place the alidade at right angles to its first position upon the board, repeat the operation, clamp the large screw again, and the table is level. Now free the tangent screw by loosening its clamp, place the edge of the rule *r* upon the occupied point *s* and the point *b*, the telescope being directed toward the spire B, as shown by the arrow-head in the figure, and revolve the table horizontally about its center with the hands until B is seen in the field of the telescope; clamp the tangent screw and turn it till the intersection of the cross-hairs bisect the top or center of the spire B. The table is now "in position" if the plotted points be correct and the proper objects sighted. In other words, the table is "oriented" when the point observed upon and the point occupied are in the line of sight, the edge of the rule being upon the two plotted points; the one, *s*, perpendicularly over the occupied station, and the other, *b*, the station observed upon. As a test of the correctness of this, place the rule upon the point *s* again, and upon the points *a* and *c* consecutively, and if the two signals A and C are covered by the vertical cross-hair of the telescope, the orientation is assured, and the meridian of the sheet is parallel to that of the earth, all the lines joining the signals and their respective projections being also parallel.

It will sometimes happen that the upper metal disk attached to the table, after it has been clamped and the tangent screw used to put the table in position, has a tendency to spring still further with a sudden movement or slight jerk, and this movement may not occur until impelled by the ordinary working about the table, and pass unobserved by the operator. This may arise from the two disks being screwed too closely together, and the faces in contact not being sufficiently oiled. It is often the source of much trouble to the beginner, and he is unable to discover the cause. It is well, therefore, in orienting the table, when this is suspected, to take hold of the edge of the board with the thumb and finger and spring it very slightly from side to side, in order that the table may settle itself in a fixed position. The cause of the trouble must, of course, be removed on the first opportunity.

The next operation is to "take the forward line" to the next point which it is desirable to occupy or determine, either some natural object which can be occupied, or a forward signal placed for that purpose, say the signal D.

The edge of the rule is placed upon the point *s* and moved about that point as a center until the forward signal D is covered by the vertical hair, and then a line, *f*, is drawn along the edge of the rule from *s* sufficiently far to reach the estimated distance on the sheet of the point *d*, and at each end of the rule the short check lines *n n* are drawn. In the same manner lines to be afterward intersected should be drawn to such objects as it may be well to determine. To prevent confusion, the ends of such lines are marked as in the diagram: *ch.*, chimney; *t.*, tree; *cup.*, cupola; *sp.*, spire; *w. m.*, wind-mill, &c. Tangent lines, and lines radiating to objects comparatively near at hand, to be chained or obtained by the telemeter, as fence corners, &c., should be likewise taken. If the station occupied be in an elevated and prominent position, its height should be observed, both as a guide for putting in the contours at the point and to serve as a point of reference in taking heights at other places, the method of doing which will be given hereafter. The necessary sketching is now done, omitting nothing that can be completed from this point; the alidade removed, the table raised, the signal put up, and the party leaves for the next station. Sometimes it is necessary to start the chain from the station to the forward signal.

When moving from one station to another, it is the custom with some topographers to loosen the tangent clamp, with the idea that if the table come in contact with any object while being transported it will revolve and be less liable to injury. This is perhaps true, if the blow comes on the side of the table in the direction of its plane.

Fig. 2.

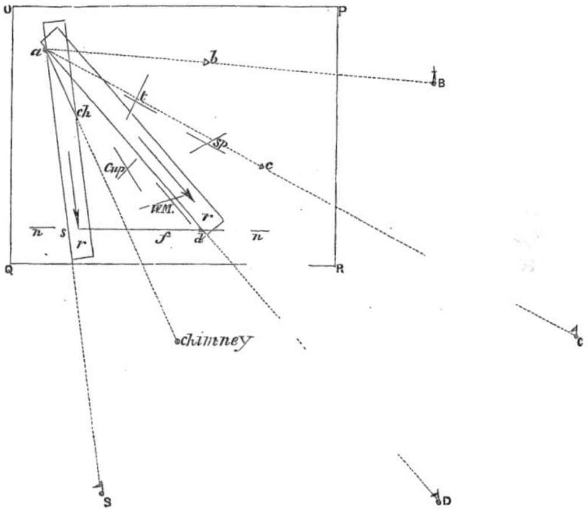

Now let the letters in Fig. 2 be the same as in Fig. 1. The table is removed to the station A and placed over the point on the ground, put in approximate position, leveled, clamped, and loosened at the tangent screw, as at station S. The rule is then placed upon the line *as*, the cross-hairs of the telescope directed toward the signal S, and the table brought into position, as before described. Then, keeping the edge of the rule upon *a*, direct the telescope upon the signal D, and draw the line *ad*, intersecting *f*, and determining the position of the point *d* upon the sheet, corresponding to D, and bearing the same relation in position and distance to the points *s*, *a*, *b*, and *c*, as the signal D does to S, A, B, and C. All the other objects to which lines were drawn from *s*, and which can be seen from A, are intersected and determined in the same manner. This is an example of the method of "intersection."

The necessary sketching, determination of height, &c., are executed here as at S, and, indeed, at every point occupied, it being desirable, if possible, never to occupy a station more than once.

The intersection of two lines is not, however, positive evidence of the correct determination of a point. Let us, therefore, proceed to D and again determine it by "resection" from the point B. (See Fig. 3.)

Fig. 3.

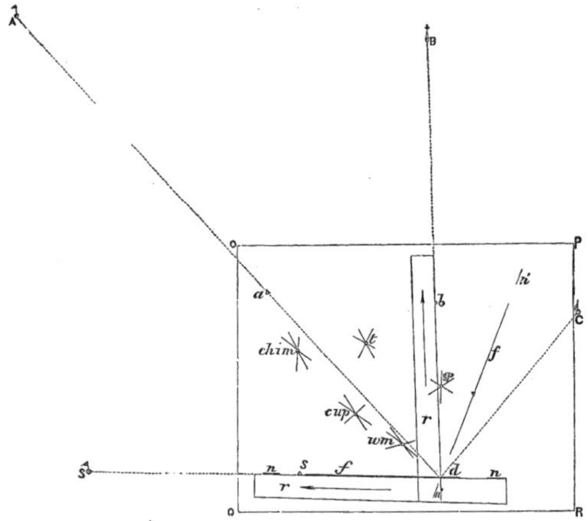

The table is placed over the point D, put in approximate position, leveled, &c., as at the other stations. The rule is then placed upon the forward line, *f*, (now called the "back line," as seen from D,) passing through the point *s*, so that the checks *n n* are just visible along the edge, and the telescope directed toward the signal S, as shown by the arrow, and the table oriented. The rule is then placed with its edge bisecting one of the plotted points, such as *b*, which will give a cleanly cut angle (the nearer 90° the better) with the line *f*, and is moved about that point as a center until the spire B is covered with the vertical hair. A line is now drawn along the edge of the rule, crossing the line *f*. If this line intersects *f* at precisely the same point as the lines *f*, *a*, and *d*, the position of *d* upon the map is assured, and a delicate hole with the dividers should be pricked upon the sheet to fix the point, surrounded by a small circle in pencil. The point may be still further tested by resection from C. If the forward line from *s* has also been chained, the distance taken from the scale and laid off from *s* on the line *f* will afford still another test, and it is quite sufficient if it agree with an intersection where only one can be obtained.

Another forward line, *f'*, is now taken, with the usual checks, *n' n'*, to the next desirable station, and lines of intersection are also drawn from *d* upon the chimney, wind-mill, cupola, tree, and spire previously observed, as they appear in the telescope, in succession from left to right, and their positions definitely fixed upon the map, pricked through and marked; and these being well determined, can now be used for the determination of other stations. New lines to such other objects as may be thought necessary should be taken, as well as tangent lines, and then follows the sketching to fill in the details about the station.

Fig. 4.

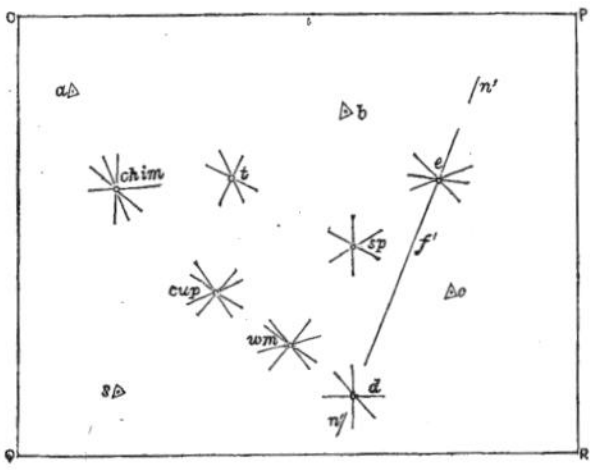

The table is now removed to E, Fig. 4, (which it was thought unnecessary to mark on Fig. 3,) through which the forward line from *d* is supposed to pass, and is placed over the station; and the point *e*, representing the projection of the signal E upon the map, is determined by resections by the use of the line *f'* and the points *s*, *a*, *b*, and *c*, although the latter two are not absolutely necessary. The spire and tree may also be used for this purpose. Those points which, owing to acute intersections, have been insufficiently determined, as the chimney, cupola, and wind-mill, are again intersected. Other intersecting lines are taken from *e* upon other points which present themselves, the necessary sketching made, and a new forward line taken to the next station.

During all these operations occasional recurrence should be had with the alidade to some established point to assure the immobility of the table, or to correct any deflection from the true position which may have taken place.

If upon going to a forward signal or object to which a line has been taken it is found that it cannot be occupied, or that it is in such a position that a sufficient number of points cannot be seen from it, or, for any reason, it does not answer the desired purpose, a point in range between the two stations, or upon the prolongation of the line connecting them, can be occupied. Getting into line between the two stations is performed by two persons standing facing each other, about thirty meters apart, and as nearly on the line as possible, one of whom sees the back and the other the forward signal. Each then moves alternately to the right or left, as directed by the other, until each signal is in line with the person whose back is toward it, as seen by the person facing it. The table can then be readily placed anywhere on the line. A position for a point beyond the forward signal may also be found by simple alignment with the two signals.

When by accident in drawing a forward line from an occupied point, near which upon the sheet is plotted another or several other points, the rule is not set upon the point occupied, and the error is not manifest until the forward signal is reached, instead of going back to take the line over and draw it from the last station, it can be constructed by drawing from the correct station a parallel to the false line.

Points and lines.—The accuracy of the work on the topographical sheet is primarily and mainly dependent upon the correct determination of points, and a want of an exact knowledge of the capabilities of known points to determine the observer's position anywhere upon the sheet, as well as the positions of the other points, is one of the greatest sources of trouble and error to a beginner. When a survey is commenced with slightly faulty points, and uncompensated as the work proceeds, the scale upon which it is executed becomes variable, and consequently erroneous.

When, as we have seen, a triangulation point is occupied, and lines drawn from a number of other plotted points with the table in position intersect perfectly at that point, the position is assured; but when they do not thus intersect, the cause of the difficulty may be found either, 1st, in errors of triangulation or computation; 2d, in a faulty projection or plotting of the points; or, 3d, in the unequal expansion or contraction of the paper. The first two, when at all great, can only be corrected by a revision of the work of triangulation and projection; and the latter, if not sufficiently large to warrant an entire rejection of the sheet, can be remedied only by the judicious ction of the topographer, with the plane-table, in the field.

When the points disagree within quite moderate limits, an experienced topographer can, by distributing the error among the points in the proportion of their distances from the occupied point, so reduce the effect of the sum of these distributed errors on the position of the occupied point that he may be safe in considering positions determined from his point, so corrected, as more accurate and trustworthy than the plotted points themselves, and use them as such. A maximum error of twenty meters on a scale of $\frac{1}{10000}$ can generally be reduced at the point of intersection to an almost, if not quite, imperceptible quantity.

The topographer should be slow to reject a point as unfit for use. No matter what the apparent disagreement may be, one set of points should not be hastily thrown aside and another accepted because one set *appears* to agree and the other to disagree. But the positive occupation of a series of points whose accuracy thus becomes established, and of another series whose inaccuracy is equally well determined, renders the preference of one set over the other at times not only permissible but obligatory. It should always be remembered that absolute and careful investigation in the field, and close examination of the projection, plotting, and data of triangulation ought to be made before any point or set of points is condemned.

THREE-POINT PROBLEM.

It is often expedient to set up the table in position at an undetermined point without any back line on which to set. With three signals in view whose positions are projected on the map, the table can be oriented and the point determined by means of the "three-point problem."

The table is brought into approximate position by the eye or declinatoire, and, not being properly oriented, the lines drawn from the three projected points will not intersect in one point, except when all four are on the circumference of a circle. In this case the "two-point problem" is available. Except in this instance, however, the lines will form a small triangle, called the triangle of error, or two of them will be parallel, intersected by the third. The position of the true point can then be determined geometrically from these several intersections, and is always at the point of intersection of arcs of circles drawn through each two points and the point of intersection of the lines drawn from them; but the construction of these arcs is inconvenient in the field. More practicable modes of locating the points sought will be given in their order.

In the classification given below, based upon the location of the true point in relation to the triangle of error, the triangle formed by the three fixed points is called the *great* triangle, and the circle passing through the same points the *great* circle.

CLASS 1.—When the point sought falls within the great triangle, the true point is within the triangle of error. (Case 1.)

CLASS 2.—When the point sought falls within either of the three segments of the great circle formed by the sides of the great triangle as chords, (Case 2,) or without the great circle and within the sector of the opposite angle of either angle of the great triangle, (Case 5,) the true point is on the side of the line from the middle point opposite to the intersection of the lines from the other two points. This also includes Case 3, where the three fixed points are in a straight line, in which case the points are considered as being in the circumference of a circle of infinite diameter, and the true point always lying in one of the segments of the great circle.

CLASS 3.—When the point sought falls without the great circle and within the sector of either angle of the great triangle, the true point is on the same side of the line from the middle point as the intersection of the lines from the other two points. (Case 4.)

In case the point sought falls on the range of any two of the points, and the table is deflected from true position, the lines from the two points will be parallel, intersected by the line from the third point. But this range can always be determined by alignment, the table set in position on the range, and the point occupied determined by resection on the third point. (Case 6.)

In case the point sought falls *near* the range of any two of the three points, the lines from the two points are so nearly parallel that their intersection falls off the table, but the relation of the true point to the triangle of error is in no way changed.

The accompanying diagram shows the fields embraced by the classes given above, also the location of each of the cases included in those classes.

A point on the circumference of the great circle being indeterminate, it is apparent that a determination should not be attempted in close proximity thereto, if better conditioned points are available.

The following cases are believed to include all possible conditions of the relation of the position of an undetermined point to three fixed points. The surveyor is supposed to face his signals and the directions right and left given accordingly:

CASE 1. (Figure 1, Plate 31.)—When the point sought is within the great triangle, the true point is within the triangle of error.

a b c are the projected points, and *ab ac bc*, the false intersections from them forming the triangle of error.

Rule.—If the line from any one of the points falls to the right of the intersection of the other two, turn the table to the left, and if to the left, turn it to the right.

When the point sought is without the great triangle the true point is also without the triangle of error, and is situated to the *right* or *left* of it, according as the table is out of position to the *left* or *right*.

CASE 2. (Figure 2, Plate 31.)—When the point sought is without the great triangle and within the great circle, the true point is without the triangle of error, and the line drawn from the middle point lies between the true point and the intersection of the other two lines. This also includes Case 3, (Figure 3, Plate 31,) which rarely occurs in practice, where the three points are in a straight line.

Rule.—If the line from the middle point is to the right of the intersection of the other two, turn the table to the right, and if to the left, turn it to the left.

CASE 4. (Figure 4, Plate 31.)—When the point sought is without the great circle, and the middle point is on the far side of the line joining the other two points, the true point is without the triangle of error, and upon the same side of the line from the middle point as the intersection of the other two lines.

Rule.—If the line from the middle point is to the right of the intersection of the other two, turn the table to the left, and if to the left, turn it to the right.

CASE 5. (Figure 5, Plate 31.)—When the point sought is without the great circle, and the middle point is on the near side of the line joining the other two points, the true point is without the triangle of error, and the line drawn from the middle point lies between the true point and the intersection of the other two lines.

Rule.—If the line from the middle point is to the right of the intersection of the other two, turn the table to the right, and if to the left, turn it to the left.

CASE 6. (Figure 6, Plate 31.)—When the point sought is on the range of either two points, and the table deflected from true position, the lines drawn from these points will not intersect, but will be parallel, intersected by the line drawn from the third point; the true point is then between the two parallel lines.

Rule.—When the line from the right-hand station is uppermost, turn the table to the right, and when that from the left hand is uppermost, turn it to the left.

Practicable modes of determining the position of a fourth point by resection upon three fixed points.

1st. *Lehmann's method.* (Figure 7, Plate 31.)—This method is based upon the fact that the point sought is always distant from the three lines drawn from the three fixed points in proportion to the distances of the latter from the point occupied.

A B C are the projections of the three signals from which it is desired to determine by resection the position of a fourth point D. The table being out of position to the right, the triangle of error formed by the three lines from A B and C is *ab ac bc*. The true point occupied lies at D, being at the intersection of the circles AB *ab*, AC *ac*, BC *bc*. Now, if perpendiculars be drawn from D to the lines drawn from A B and C, we shall have

$$Da : Db :: DA : DB, \text{ or } Db : Dc :: DB : DC.$$

The relative distances of the point occupied from the three signals must be estimated and the point located in reference to the three lines from A B and C accordingly.

Netto's method. (Figures 8, 9, and 10, Plate 31.)—This method of determining the true position from the false intersections is ingenious and of much practical value.

The table not being properly oriented, and having resected upon *a b* and *c*, we have the triangle of error $ee'e''$. Now, by the Lehmann method, judge of the position *d*, (the point sought.) Set the alidade on *db* and revolve the table so that the line of sight passes the signal B. Resect again on *a b* and *c*, and we have the triangle of error $gg'g''$. Join *e* and *g*, and through both points draw parallel lines *ii* and *kk*. Lay off $ei=ef$ and $gk=gh$. Join *i* and *k*, and the intersection *l* lies in the line of sight from the true point to the middle point *b*. Set on this line, resect upon *a* and *c*, and *d* is the point sought.

If the two triangles of error are situated on the same side of the true line of sight to the middle point, the parallel lines are set off on one side of *eg* only.

The triangles of error $ee'e''$ and $gg'g''$ are always similar, $\angle g''=\angle e'$, $\angle g'=\angle e''$, $\angle g=\angle e$, and as the two points *e* and *g* are always in the circumference of the same circle, if the table is deflected equally on the opposite sides of the true line of sight to the middle point, the triangles of error will be equal and $ef=gh$. On the true line of sight *gh* and $ef=0$.

In the triangles *gkl* and *eil*, *ii* and *kk* being parallel, $\angle g=\angle e$, $\angle l$ is common, therefore $\angle k=\angle i$, and the triangles are similar, and $el : gl :: ei(=ef) : gk(=gh)$.

(Figure 9, Plate 31.)—The point sought (*d*) must lie in the circle passing through *aec*, and also through *agc*. Draw the circle *agdecs*, join *s* with *e* and *g*, then we have

$$\angle dse=\angle dce \text{ and } \angle dsg=\angle dag$$
$$\angle dce=\angle dbe'' \text{ and } \angle dag=\angle dbg''$$
$$\text{Therefore } \angle dse=\angle dbe'' \text{ and } \angle dsg=\angle dbg''$$

also *se* parallel to bc'' and *sg* parallel to bg''
and the triangles *sle* and *blf* are similar,
and the triangles *slg* and *blh* are similar;

from which we get $le : lf :: ls : lb$

and $lg : lh :: ls : lb$,

also $le : lf :: lg : lh$ and $le : le - lf :: lg : lg - lh$

that is $el : ef :: gl : gh$, or $el : gl :: ef : gh$.

The amount of the angle at l is always an indication of the value of the determination of the nt sought. The more obtuse the angle the better the determination.

BESSEL'S METHODS.

Bessel gives two methods, both based on the same principle.

First method.—(Figures 11, 12, and 13, Plate 31.)

Let a b and c be the projections of the three points observed upon, and ab bc ac, the triangle of error formed by resection upon them when the table is not in position. Lay off bc' on $ba = bc$, xtend bc and lay off $ba' = ba$. Call the angle at the intersection $ab = x$, and that at the intersection $bc = y$. At a' lay off toward you $\angle ba'e = \angle y$, and at c' in the same direction $\angle bc'e = \angle x$. The lines o laid off will intersect in e, which lies in the line of sight through the middle point b and the point sought, (d.) By resection upon a and c, the position of the point on this line is fixed.

The solution of this is as follows, (Figure 14, Plate 31): Lay off at a the angle $bae = \angle bdc$, and at c $\angle bce = \angle bda$, drawing the line be, $\angle ebc = \angle abd$ and $\angle eba = \angle cbd$. Produce bc to f, so that $bf = ba$, and draw fg parallel to ce. Lay bfg on ba, so that f falls on a and g on h.

Then we have in the quadrilaterals $ahbe$ and $abcd$

$$\angle bae = \angle bdc, \quad \angle hbe = \angle abc$$
$$hb : be :: bg : be :: bf : bc :: ba : bc.$$

The two quadrilaterals are therefore similar, and hence

$$\angle ebc = \angle hba = \angle abd$$
$$\text{and } \angle eba = \angle cbd.$$

Second method.—The plane-table may also be put in position without the use of the points a' and c'. (Figures 15, 16, and 17, Plate 31.)

On ac at c lay off $\angle ace = \angle x$, and at a lay off $\angle cae = \angle y$. The lines so laid off will intersect in e, which lies in the true line of sight through the middle point b and the point sought, (d.) Resection upon a and c then fixes the position of d.

The angles e and d of the quadrilateral $aecd$ are equal by construction to two right angles; hence a circle may be described about the quadrilateral, and we have the periphery angles

$$ace = adb$$
$$\text{and } cae = cdb.$$

This latter method, being simpler, is better than the first, but, under certain circumstances, one may be used when the other cannot. If, for instance, by the last-mentioned manner of construction, the point of intersection (e) should fall outside the plane-table, it may possibly be made to fall inside by the first method. Again, if, by the latter method, the angles of intersections happen to be right angles, or nearly so, then the two plotted lines to e become parallel to each other, or nearly so, in which case the first method may be used with advantage.

The best mode of constructing the angles x and y upon ac is with the alidade; directing the line to one of the objects and observing the other object with the alidade set upon the point at which the angle is to be set off. It can also be readily done with the dividers by laying off the chord of the angle.

Should either or both of the angles set off at *a* and *c* be so obtuse that the point *e* falls off the table, a shorter base can be used, drawn parallel to *ac*, as near to *b* as may be necessary.

TWO-POINT PROBLEM.

The occasion may arise where it is desirable to place the table in position at a given point, from which point only two determined points are visible. This may be done by the following methods. The first mode possesses the virtue of making no linear measurement, and demonstrates in a very satisfactory manner the power of the table in determining position by resection. (Figures 18, 19, 20, and 21, Plate 31.)

Two points, A B, not conveniently accessible, being given by their projections *a b*, to put the plane-table in position at a third point C. (The capital letters refer to points on the ground and the small ones to their corresponding projections.)

Select a fourth point D, such that the intersections from C and D upon A and B make sufficiently large angles for good determinations. Put the table approximately in position at D, by estimation or by compass, and draw the lines A*a* B*b*, intersecting in *d*; through *d* draw a line directed to C. Then set up at C, and assuming the point *c* on the line *d* C, at an estimated distance from *d*, and putting the table in a position parallel to that which is occupied at D, by means of the line *cd*, draw the lines from *c* to A and from *c* to B. These will intersect the lines *d* A, *d* B at points *a′* and *b′*, which form with *c* and *d* a quadrilateral *similar* to the true one, but erroneous in size and position.

The angle which the lines *ab* and *a′b′* make with each other is the error in position· By constructing now through *c* a line *cd′* making the same angle with *cd* as that which *ab* makes with *a′b′*, and directing this line *cd′* to D, the table will be brought into position, and the true point *c* can be found by the intersections of *a*A and *b*B.

Instead of transferring the angle of error by construction, we may conveniently proceed as follows, observing that the angle which the line *a′b′* makes with *ab* is the error in the position of the table. As the table now stands *a′b′* is parallel with A B, but we want to turn it so that *ab* shall be parallel to the same. If we, therefore, place the alidade on *a′b′* and set up a mark in that direction, then place the alidade on *ab* and turn the table until it again points to the mark, then *ab* will be parallel to A B, and the table is in position.

Another method is as follows: (Figure 22, Plate 31.)

Two points, A and B, not conveniently accessible, being given, to put the plane-table in position at a third (undetermined) point, C.

Set up the table at the point sought as nearly in position as can be done with the eye, and resect upon A and B, intersecting the line *bc* at *c′*. The angle *ac′b* is the true angle at the point occupied, subtended by A B, being the angle of nature actually drawn; therefore, the true point must be on the circumference of the circle passing through *abc′*. Construct this circle. Measure off a base, C D, at least half the length of C B, at right angles, or nearly so, to *bc*, in either direction most convenient. Set up a signal at D, and with the alidade draw the line *c′d*. Remove the table to D, and, by means of a signal at C, (the point sought,) and the line *dc′*, bring the table into a position parallel to that which is occupied at C. With the alidade centering on *d*, observe the signal B, and draw the line *db′* intersecting *cb* at *b′*. *c′b′* is the distance of the point C from B, and this distance laid off on the circle *ac′b* as a chord from *b* will give *c″*, the true position of the point C. A fourth point may then be occupied, and by resection upon A B and C the accuracy of the determination of C verified.

Where it is possible to get the two signals A and B in range, it is easy to determine the position of a third point by a mode long practiced by topographers.

Set up the table anywhere on the range line, and, having set up a signal at the point sought, resect upon it, intersecting the range line anywhere, and, by means of the range signal and the line to it, the table may be set in parallel position to that occupied in the range, which is the true position, and the point sought may be determined by resection upon the two fixed points and their projections.

FIELD-WORK.

In taking lines of intersection upon a point or object from a series of stations, when these lines do not coincide in one point, as they are usually derived from stations at unequal distances from the point, the error should not be divided equally among them, but in proportion to their lengths.

It should be borne in mind that very short lines from a determined point—as, for instance, to the corners of a fenced road, where the table occupied the center of the intersection of two roads—may be taken with no apparent error when the table is deflected to some extent from its true azimuth, but that in this case a prolonged line will be considerably out at its further extremity.

A long line should never be obtained by the prolongation of a short one from a back station where there is no small check line, or some other point in that prolongation already fixed.

It will be apparent that the more nearly at right angles intersecting lines cross each other, the more clearly the point will be defined; acute intersections, as far as possible, should be avoided, and, even when they are crossed by a third line at a satisfactory angle, a fourth line, or an accurately chained distance from a well-determined point, is advisable.

The necessity for dependence upon a measured line, with an established direction alone, for position is sometimes unavoidable; but, except for minor details, it should never be resorted to when other means are available, and, in finished work, no lengthened consecutive series of chained lines for positions should be trusted without resections for tests of accuracy. It is safer to combine both, unless the supply of signals is ample and they are favorably located.

A judicious use of range lines from established points, or signals, will much economize time and facilitate work.

Two range lines from well-determined points are equivalent in value to four intersecting lines.

Tangent lines can only be used for determining the edges of woods, bends of streams, sweeps of shore-line, outlines of shoals, small ponds, and the outlines of other objects when in unimportant localities, and are inadmissible for any purpose in which accuracy of delineation is required, save when they form a polygon, by which absolute convexity at all points is to be represented, and even then the points or objects should be visited and sketched, if possible.

Where the topography surveyed includes the shore-line of a body of water, and immediately precedes the hydrographic survey thereof, as in the Coast Survey, it is the duty of the topographer to locate and determine the shore signals, and they should be placed so as to furnish the hydrographic party with the best points available for the determination of positions on the water.

It is well to mark all stations occupied along or near the shore by pegs driven into the ground with stones about them, so that their positions may be found by the hydrographer if the signals should be destroyed, and he can then select such as are best adapted to his use.

Natural or artificial objects along the shore, or in plain sight from the water, such as fence ends, rocks, prominent houses, and posts on wharves, &c., should be determined and marked upon the sheet. As some time may elapse between the labors of the two parties, the stations should be well secured above the wash of the tides.

Lines to buoys and other permanent floating objects should be, as far as practicable, taken at the same stage of the tide, or direction of current, or the status of the tide noted at the time of observation.

In the determination and tracing upon the chart of the low-water line, so much in its outline is generally dependent upon the direction and force of the wind that no fixed rules for guidance can be given.

The delineation of the ordinary mean low-water mark should be aimed at, and when it is beyond the reach of the plane-table, and presents no marked points for determination, or is of a character that will not admit of putting up and working the instrument—as along swampy shores of the South, where the muddy shoals extend far sea-ward, and among the shifting quicksands of our great estuaries and bays—it must be left to be traced by the soundings and tidal reductions of the hydrographic parties.

It is always best to determine the high and low water lines, both at spring and neap tides. Having learned the range of tide, the topographer will know how long he can work without error.

Where, on the occurrence of any great or unusual storm or freshet during the working season, the low-water line, which has already been surveyed, is found to have changed in form or locality, it should be resurveyed, and both the old and new outlines retained on the sheet, with the appropriate notes.

As a feature which is quite interesting and important under certain circumstances to the hydrographer, low-water springs, having their origin and outlet below the high-water line, should be shown on the chart, where it can be done, in the regular routine of work. All grassy shoals should be delineated. They are always found in water scarcely agitated by waves or currents, and their shape and outline on the channel side is a very marked feature, and a good measure of the power of the current. Eel grasses should also be put upon the map, as indicating an antecedent accumulation of fine sand or soil.

Orientation by the declinatoire or compass-needle, alone, is not reliable, unless for obtaining positions for rough sketching in plane-table reconnoissance, but it may be useful as an adjunct when an operator, in default of sufficient points, desires to obtain an approximate position. It is used by placing the straight edge of the box containing the needle upon a magnetic meridian, previously traced upon the chart, and revolving the table until the needle points to 0°, or north, on the graduated arc in the end of the box. The magnetic meridian is roughly obtained at any well-determined station, when the table is properly oriented by the use of the declinatoire itself, the meridian line being drawn upon the sheet along the straight edge of the box when the needle points to 0°. When the declinatoire is fastened to the rule of the alidade, the line is drawn along the edge of the rule.

In sketching or drawing, care should be taken not to lessen the size of natural objects, the scale being followed as far as practicable; but in some cases, which will be apparent, it may be desirable to enlarge somewhat, but very cautiously. The topographer should learn to draw from nature readily, and at once, without being obliged to erase or interlineate.

By working carefully at first, facility in this will be arrived at in due time.

Too frequent use of India-rubber disturbs the fibers of the paper and renders the subsequent inking less neat and clear. The drawing should be plain and distinct, so as not to be obliterated easily by the movement of the alidade over the paper, but not so dark or heavy as to blur. The object should be solely to represent accurately the surface and elevation of the country surveyed, and it should be easy and natural, and not a stiff copy of conventional signs.

As far as practicable all work should be drawn in the field under the eye. Sketching and plotting in the office from notes is objectionable, unless the country be near at hand for examination in case of doubt or a defective sketch or of error of chaining.

Too great care cannot be taken in the manipulation of the plane-table. There should be no pressure, and in moving the alidade both it and the arm should always be raised clear of the board, so as not to rub over the surface of the sheet.

The topographer should learn to distinguish, as a matter of economy in point of time, between the relative importance of different topographical features. While it should be the object to do all the work correctly, yet a discrimination should be made between the expenditure of time necessary to a correct representation of the thickly-settled streets of a town and the bend of a small creek in an obscure and unimportant swamp.

CONTOURS.

If there be any feature which more peculiarly distinguishes one section of a country from another, imparting to it its most striking characteristic, and to which all other accidents of ground are subordinate, it is to be found in the inequalities or changes of the elevation of the surface, and it is the correct representation of this feature that calls forth the best skill and judgment of the topographer, and upon which the value of a map most materially depends.

Previous to the commencement of work, it is well to become possessed of some knowledge of the country which is to be surveyed. A rapid examination of the ground, by showing whether it lies in regular parallel ridges and intervening valleys, isolated hills, gradually sloping plains, or broken, abrupt, and rocky declivities, and whether partially or entirely open or wooded, will suggest to the topographer the best method of operation, and enable him to form a general plan of work which will result in economy in point of time, and as a consequence will preserve the continuity of contour from day to day as the work progresses.

By keeping in view the characteristic features of the hills in the section of country under survey, the topographer will be able to give a naturalness to a sheet, which by a mere formal delineation by prescribed rules he could not obtain. It is to be observed that the elevations and depressions, in their form and course, follow, or rather are a part of, a general system of nature, however capriciously detached localities, or even extended areas, may appear to be excepted from the general law. Where these exceptions are found they can usually be traced to some breach of or interference with this law, and will be found to be confined solely to the locality where the disturbing causes operated.

Thus the principal ridges will be found to tend in one general direction, either running parallel with a main range which lies further off, or forming spurs at right angles to it. Along the coast the latter is generally the case, as we there usually find the spurs, or the extremity of some main prominent inland range. When a single detached hill is found, or a series of them, presenting, as is sometimes the case, so smooth and regular an outline as to be compared to half a watermelon, and apparently located without any reference either in direction or character to the other elevations of the vicinity; and so also when the ground presents the appearance of a confused mass of broken bluffs, rocky faces, and cleft surfaces thrown together without any apparent regard to order or regularity, it will be found that the general delineation, when followed far enough, will show that the contours, whatever may be their local complexity and irregularity of outline, follow the same general direction as the main ridges of the sheet. Along the intermediate shores, upon islands, or long arms running far out into the sea, where the sandy knolls, or dunes, and ridges, shift or change their outline under the influence of wind and tide, an exception is found which calls for careful delineation, and the peculiar and striking character and forms of these should be portrayed with all possible exactness as an interesting and useful aid in the study of a correlative branch of geodesy.

To the meteorologist and to the physical geographer the careful mapping of dunes is valuable; for however familiar the locality may be, no eye-view can discover those recurring features which are found in the map of an extended district.

If a sandy district is exposed to permanent or prevalent dry winds, traveling dunes will be found. These are distinguished from other hills by the contrast between the *fore slope*, (on the lee side,) which is steep, and the near slope, (on the windward side,) which is gentle. Successive surveys on sandy coasts, where close attention has been given to the contours of the dunes, are of great value for comparison if every detail is carefully given. Even in places where the winds are uncertain, proper contouring discovers dunes which travel along the resultant of the forces, and in the direction of this resultant the great dip of the fore slope is found. For such dunes the several slopes for different points of the compass are equally interesting.

On exposed points projecting far into the sea, peculiar dunes, called *galls*, are found. They are long ridges of sand, broken by slough-ways. It has been observed that sometimes these slough-ways are parallel. Correct and well-contoured topographical maps of the localities where they are found would aid much in a study of this interesting subject. Contours only would discover their order and exhibit the material system, which could not be done by a representation by hachures.

The delineation of bluffs along the shore should also, for like reasons, be carefully executed, and when it is possible a representation of their slopes should be given. Bluffs, if worn by the waves, will usually exhibit three slopes: 1st, the caving slope at the top; 2d, the talus; 3d, the apron or flat, exposed wholly only at very low tides. The caving slope is sometimes perdendicular where tertiary country is being worn away—never where old dunes are yielding. The talus is usually of selected material—stones, perhaps. The talus and flat are generally wanting in bluffs worn by currents.

As has been said, in no branch of surveying does so much depend upon skill, combined with good judgment and experience, as the faithful representation of hills over an extended and diversified area, and long practice and close observation only can give facility and accuracy in its execution.

Various methods, more or less defective in presenting a correct idea of elevations and depressions, have been contrived for topographical surveys; but the graphic representation of the successive gradations of level by means of horizontal lines, as at present employed in the Coast Survey, gives the nearest approximation to nature which has yet been devised, and when faithfully executed must necessarily express very nearly, if not exactly, the shape and height desired.

Contours, or horizontal curves, are the outlines of horizontal sections of ground at different elevations, with designated equal intervals between their planes, delineated in their true positions relatively to each other and the rest of the map, and agreeably to the scale of the map itself; or, briefly, a contour is the curve produced by the intersection of a horizontal plane with the surface of the ground.

Perhaps contours may be described more simply as imaginary shore lines formed at stated or regular elevations by the water which is supposed to rise successively to these elevations over the face of the country.

As each curve has equal vertical ordinates at all points, the elevation or profile of a hill, as well as a model in relief, can be constructed directly from the map, when it is accurately executed on a large scale, without further field measurements.

A profile of a hill is the outline or trace formed with its surface by a vertical plane cutting the hill in any direction.

The annexed diagram, from actual survey, shows the profile, through the line A′B′, of the hill H, as represented on a topographical map. The full parallel lines upon the profile represent lines of equal elevation for every twenty feet difference of level, and the broken or intermediate lines *x x x*, those of ten feet.

A reference to the letters upon the diagram is all that is necessary for a full understanding of the subject; *a* is the shore line or high-water mark upon the map, *x x x* are the auxiliary ten-feet curves, *f′* the coincidence of curves upon the chart at the perpendicular face of the hill, *f* upon the section. This is the only case where contours of different heights run into each other upon a topographical plan. D′ D′ are depressions in the face of the hill, represented on the profile by D D; *d′* is a barranca or dry broken gully, and *c′ c′* a water-course.

It will be plain that if we were to suppose the water to rise to a height of twenty feet above the high-water line, or to *h* on the profile, the twenty-feet curve upon the map would then become the shore line, and the depression D′ would become a pond of water; and if the water were to rise to a height of thirty feet, the dotted broken line would form the shore line, and the knoll G would become an island.

Horizontal curves are drawn upon the map with the eye, after having obtained the elevations by means of which their positions are fixed, by the measurements of vertical angles with the arc of the alidade, or in detailed or special surveys with the level. Where the slope is regular and tolerably steep, the tracing of them is attended with but little difficulty; but where the rise is very gradual, giving large horizontal distances between the contours, even when the vision is unobstructed, or where the country is much broken or thrown into irregularly-shaped knolls and depressions, presenting an intricate and confused variation of surface, the correct representation is, at times, very perplexing.

As in some instances, owing to the smallness of the angle to be measured, the vertical arc cannot be relied upon for close determination of heights, and it is evident that the nearer to a level a country is, the nearer it is necessary to obtain the exact elevation for the location of the contours, recourse to the leveling instrument is indispensable. With the beginner the observations for elevation can hardly be too frequent, and he should constantly bear this in mind while at work, as well as the necessity for leaving frequent well-marked points of reference wherever they may be of value for determination of subsequent heights.

When triangulation points are occupied, or positions are determined by the plane-table preliminary to the regular work; in fact, whenever any station is occupied, its elevation should be determined, as well as the heights of such other prominent points or objects as may be visible from it. Then in using these in working by resection, the topographer has as many points of reference for the determination of height as he has for determining his position.

It is well, also, to get observations for heights as often as possible upon or from the plane of reference or high-water mark; and advantage should be taken, whenever the opportunity occurs, of observing from the shore line in wooded districts upon any hill tops or conspicuous objects exposed by openings, and also upon rocks or any other natural or artificial objects upon the sides of the surrounding hills.

Where the heights of certain points have been determined by the triangulation party, or a few prominent ones have been determined with special care by the theodolite or level, they should be used as often as possible as points of reference.

When hills are inaccessible, the determination from accessible points of both the positions and heights of objects upon them, such as trees, rocks, stumps, &c., from which the positions and courses of the contours can be determined, should be sought for, and in thick forests the roads, paths, or the dry beds of streams may be found available for the use of either the vertical arc or level.

Under certain conditions, as those of densely wooded heights and vales, inaccessible, precipitous ledges and bluffs, &c., the operations of the surveyor are limited to an almost entire dependence upon the eye alone. This cannot be relied upon as accurate, and should be avoided as much as possible.

It is customary to represent on the usual Coast Survey field sheets of $\frac{1}{10000}$ scale contours of successive heights of twenty feet; but occasionally, owing to marked accidents of ground, it is deemed advisable to insert intermediate or auxiliary curves to develop the forms lying between them. On a scale of $\frac{1}{1200}$, contours for every three or five feet difference of level are frequently given.

Contours should be filled in from each station while carrying on the regular work, when it is possible to ascertain heights for that purpose; and in selecting positions for forward signals, and in prosecuting the work, reference should be had to the continuous and successive tracing of the contours on the map, both of those which are filled in *at* and those *from* the station, as this will generally prove an economy of time, and the work can be executed with more facility. When the topographer is at a station whose elevation has been accurately determined, lines should be drawn to objects of equal height in all directions and so marked, subsequent intersections giving their positions; this will be found of great assistance in running contours and as checks on heights by vertical angles. Care should be taken in sighting to distant objects to allow for the curvature of the earth.

With regard to the method of putting in contours from the base of the hill upward, or *vice versa*, opinions differ, some preferring one course, some another; but it makes but little difference, if the height of the point of beginning be correct. Neither of these systems should be specially adopted, but the peculiarities and demands of the other topographical features should be considered, and all should be worked together to the best advantage, so as to make the work at each station as complete as possible before leaving it.

When, as often happens, the work has been carried on in a wooded country, in a place where observations for height have not been made for a long time, then the importance of coming out upon some point of known elevation is evident.

If this is impossible, as is sometimes the case, as in the filling in of the topography in dense woods, in a rolling country, where the topographer is confined to the roads, frequently with very short sights, and there is no check to come out upon, or when the work closes on the edge of the sheet, the use of the level in the hands of an aid, though consuming time, is indispensable.

The determination of the heights of artificial features, such as fence corners, houses, &c., as an assistance in contouring, should not be neglected.

When the contour runs very near any remarkable accident of ground, as a prominent spur or indentation, on general field maps of $\frac{1}{10000}$ scale, a slight deviation above or below its true plane is admissible, although it is preferable to represent it by the introduction of the auxiliary curve, as shown in the sketch.

It is very desirable that all features within the twenty feet curves, such as breaks in the ground, isolated boulders, rocks, &c., which cannot be legitimately represented by auxiliary curves, should be shown by hachures or conventional signs. When the rocks have a distinct stratification, or when cleft in certain directions, it should be indicated, if they or the scale of the map be sufficiently large to warrant it.

It may happen that small features, which are unimportant in themselves, may interfere with the development of the general form of the contours, or their introduction may tend to produce confusion; these are best omitted, but this omission should be optional only with one of experience. It is dangerous to give latitude in this respect to a beginner.

When there is an abrupt rise, as in low bluffs, railroad embankments, &c., not above ten feet in height, on a scale of $\frac{1}{10000}$, it should be indicated by hachures always tapering downward, and all hachures should, in their direction, follow the downward flow of water or alluvion.

Depressions of the ground in the midst of level tracts, or upon tops or slopes of hills, unless distinguished by ponds or marsh, should be marked with the letter D in red.

The distinct summits of hills should be marked in figures when they form characteristic or remarkable features of the country.

In measuring for heights or depressions with the alidade, the plane-table is carefully leveled and firmly clamped, the telescope is directed toward the point of observation, and moved until the cross-hairs are in the same vertical plane as the observed object. The telescope is then clamped by means of the screw at the top of the arms of the vertical arc, and the central cross-hairs, at or near their intersection, are brought to cover the observed point by means of the tangent screw attached to the graduated arc. The angle read, and the distance between the occupied point and the observed point measured on the map, the height is taken from the table, which is appended.

Example.—Observations for height at an occupied plane-table station.

Stations observed.	Observed angles.	Distance in meters.	Relative height in feet of plane-table station.	Height of observ'd station in feet.	Height of plane-table in feet.	Corrected heights of ground in feet.
	° ′					
Shore signal, (angle of depression)	1 57	1,756	+ 107.4	0.0	4	193.4
Smith's Hill, (angle of elevation)	0 02	940	— 2.0	201.3	4	195.3
Black Rock, (angle of depression)	2 13	539	+ 68.7	126.9	4	191.6
Mean height of station						193.4

Detail of use of table of heights.

Shore signal:

1° 50′ for 1,700 meters .... =	179.00 feet.
1° 50′ for 50 meters = 0.1 of 500 meters .... =	5.26 feet.
1° 50′ for 6 meters = 0.01 of 600 meters .... =	.63 feet.
0° 07′ for 1,700 meters .... =	12.10 feet.
0° 07′ for 50 meters .... =	.34 feet.
0° 07′ for 6 meters .... =	.04 feet.
Total	197.37

In these observations three stations at least should be observed when practicable, and the mean adopted. On an instrument at times unavoidably subject to such rough usage as the alidade, the adjustment of the vertical arc should be frequently examined.

A material advantage in the attachment of the longitudinal level to the alidade is found in the facility by which the instrument may be used as a level in following outlines of equal elevation, making it particularly serviceable in this respect on gradually sloping ground.

A formula for computation of heights, which may prove of service where no table of heights is at hand, is appended.

When a surveyor's level is employed for all elevations, the determination of the position of the level pegs by the plane-table on the sheet may be all that is necessary, and the contours may be readily traced. The topographer should be careful, however, in this case to determine and sketch all irregular accidents of ground.

Barometric heights are admissible for approximate contours in reconnoissance where a general survey only of hills or ranges is expected.

In using the aneroid barometer in ordinary reconnoissance it will suffice to allow ninety-two feet of elevation for every 0.1 of an inch fall of the index. This allowance is for a mean temperature of the two stations of 55° Fahrenheit, and will vary with the temperature.

Leslie's formula, simple and easily remembered, is a good approximation below 2,000 feet, and convenient for aneroid observations, viz: $55{,}000 \times \frac{B-b}{B+b}=$ height in feet, B being the upper, and b the lower, reading of the aneroid. This is likewise for a mean temperature of 55°.

A very convenient instrument for a tolerably close location of contours, when carefully employed, is Locke's hand-level, which can be readily carried in the pocket, it being requisite only to know the height of the eye from the ground, and for the observer to stand equally erect at all points of observation, or to hold the level at a constant height upon a measured staff.

Formula for determining heights by a vertical angle and distance.—The difference of level consists of two parts, that which arises from the angle of elevation above the horizontal plane of the station, and that which is due to the curvature of the earth. The former depends upon the angle and distance, the latter upon the distance and the earth's radius. If a' be the angle of elevation in minutes of arc, d the distance, h the height, then, as the tangent of 1′ is $\frac{1}{3437}$, we have for the first part $h=\frac{1}{3437}a'd$, if h and d are both expressed in the same units of length; but if d is expressed in meters and h in feet, one meter being 3.28 feet, we get $h=\frac{1}{1048}a'd$. For the fraction $\frac{1}{1048}$ we may conveniently and with sufficient accuracy put $\frac{1}{1000}$ less $\frac{1}{20}$ of $\frac{1}{1000}$, and thus find the rule: *Multiply the distance in meters with the number of minutes of arc, point off the thousandth part, and subtract the twentieth part of the number thus obtained.* This will give the first portion of difference of height, whether elevation or depression.

The second term, depending on the curvature, varies as the square of the distance, and amounts to 0.22 foot in 1,000 meters, including the effect of ordinary refraction. As with the instruments under consideration extreme accuracy is not attainable, it is plain that for distances under 1,000 meters this term may be neglected. When the distance is greater we have the following rule: *Take the thousandth part of the distance in meters, square the same, having regard to the first decimal figure, and multiply by* 0.22. This term is always positive; if the first term be an elevation, it is increased; if a depression, it is diminished by the second term.

Example.—Distance = 5,500 meters; angle of elevation, 36′.

$\frac{1}{1000}d \times a'=$	198.000	$\frac{1}{1000}d$	= 5.5
subtract $\frac{1}{20}$	9.9	square	= 30.2
		multiply by	0.22
first term	188.1		
second term	6.6	second term	6.64
sum	194.7 = difference of elevation in feet.		

The above formula is near enough for distances up to ten and fifteen miles, and will not differ by as much as a foot from the result of a rigorous formula; in fact, it will keep within the limits of uncertainty of the refraction itself.

Table showing the height in feet corresponding to a given angle of elevation and a given distance in meters.

Meters.	300	400	500	600	700	800	900	1,000	1,100	1,200	1,300	1,400	1,500	1,600	1,700	1,800	1,900	2,000
Angle.	*Feet.*	*Feet.*	*Feet.*	*Feet.*	*Feet.*	*Feet.*	*Feet.*	*Feet.*	*Feet.*	*Feet.*	*Feet.*	*Feet.*	*Feet.*	*Feet.*	*Feet.*	*Feet.*	*Feet.*	*Feet.*
1′	0.3	0.4	0.6	0.6	0.8	0.9	1.0	1.2	1.3	1.5	1.7	1.8	2.0	2.2	2.3	2.5	2.7	2.8
2	0.6	0.8	1.0	1.2	1.5	1.7	1.9	2.1	2.4	2.6	2.9	3.1	3.4	3.7	3.9	4.2	4.5	4.7
3	0.9	1.2	1.5	1.8	2.2	2.5	2.8	3.1	3.4	3.8	4.2	4.4	4.8	5.3	5.6	5.9	6.3	6.6
4	1.2	1.5	2.0	2.4	2.8	3.2	3.6	4.1	4.5	4.9	5.4	5.8	6.3	6.8	7.2	7.6	8.1	8.6
5	1.5	1.9	2.4	2.9	3.5	4.0	4.5	5.0	5.5	6.1	6.6	7.1	7.7	8.3	8.8	9.4	9.9	10.5
6	1.8	2.3	2.9	3.5	4.2	4.8	5.3	5.9	6.6	7.2	7.9	8.5	9.1	9.8	10.4	11.1	11.7	12.4
7	2.1	2.7	3.4	4.1	4.8	5.5	6.2	6.9	7.6	8.4	9.1	9.8	10.6	11.4	12.1	12.8	13.5	14.3
8	2.4	3.1	3.9	4.6	5.5	6.3	7.1	7.9	8.7	9.5	10.4	11.1	12.0	12.9	13.7	14.5	15.3	16.2
9	2.7	3.5	4.4	5.2	6.2	7.0	7.9	8.8	9.7	10.7	11.6	12.5	13.4	14.4	15.3	16.2	17.2	18.1
10	2.9	3.8	4.9	5.8	6.8	7.8	8.8	9.8	10.8	11.8	12.8	13.8	14.9	15.9	16.9	17.9	19.0	20.0
11	3.2	4.2	5.3	6.4	7.5	8.6	9.6	10.7	11.8	13.0	14.1	15.1	16.3	17.5	18.6	19.7	20.8	21.9
12	3.5	4.6	5.8	6.9	8.2	9.3	10.5	11.7	12.9	14.1	15.3	16.5	17.7	19.0	20.2	21.4	22.6	23.8
13	3.8	5.0	6.3	7.5	8.8	10.1	11.4	12.6	13.9	15.2	16.6	17.8	19.2	20.5	21.8	23.1	24.4	25.7
14	4.1	5.4	6.8	8.1	9.5	10.9	12.2	13.6	15.0	16.4	17.8	19.1	20.6	22.0	23.4	24.8	26.2	27.6
15	4.4	5.7	7.2	8.6	10.2	11.6	13.1	14.5	16.0	17.5	19.0	20.5	22.0	23.6	25.0	26.5	28.0	29.5
16	4.7	6.1	7.7	9.2	10.8	12.4	13.9	15.5	17.1	18.7	20.3	21.8	23.5	25.1	26.7	28.2	29.9	31.4
17	4.9	6.5	8.2	9.8	11.5	13.1	14.8	16.5	18.1	19.8	21.5	23.1	24.9	26.6	28.3	30.0	31.7	33.4
18	5.2	6.9	8.7	10.4	12.2	13.9	15.7	17.4	19.2	21.0	22.8	24.5	26.3	28.2	29.9	31.7	33.5	35.3
19	5.5	7.3	9.1	10.9	12.8	14.7	16.5	18.4	20.2	22.1	24.0	25.8	27.7	29.7	31.5	33.4	35.3	37.2
20	5.8	7.7	9.6	11.5	13.5	15.4	17.4	19.3	21.3	23.3	25.2	27.2	29.2	31.2	33.2	35.1	37.1	39.1
21	6.1	8.0	10.1	12.1	14.2	16.2	18.2	20.3	22.3	24.4	26.5	28.5	30.6	32.7	34.8	36.8	38.9	41.0
22	6.4	8.4	10.6	12.6	14.9	17.0	19.1	21.2	23.4	25.5	27.7	29.8	32.0	34.3	36.4	38.5	40.7	42.9
23	6.7	8.8	11.1	13.2	15.5	17.7	20.0	22.2	24.4	26.7	29.0	31.2	33.5	35.8	38.0	40.3	42.5	44.8
24	6.9	9.2	11.5	13.8	16.2	18.5	20.8	23.1	25.5	27.8	30.2	32.[illegible]	34.0	37.2	39.6	42.0	44.3	46.7
25	7.2	9.6	12.0	14.4	16.9	19.3	21.7	24.1	26.5	29.0	31.4	33.8	36.3	38.8	41.3	43.7	46.2	48.6
26	7.5	9.9	12.5	14.9	17.5	20.0	22.5	25.0	27.6	30.1	32.7	35.2	37.8	40.4	42.9	45.4	48.0	50.5
27	7.8	10.3	13.0	15.5	18.2	20.8	23.4	26.0	28.6	31.3	33.9	36.5	39.2	41.9	44.5	47.1	49.8	52.4
28	8.1	10.7	13.4	16.1	18.8	21.5	24.[illegible]	26.9	29.7	32.4	35.2	37.8	40.6	43.4	46.1	48.8	51.6	54.3
29	8.4	11.1	13.9	16.7	19.5	22.3	25.1	27.9	30.7	33.6	36.4	39.2	42.1	45.0	47.8	50.6	53.4	56.2
30	8.7	11.5	14.4	17.2	20.2	23.1	26.0	28.9	31.8	34.7	37.6	40.5	43.5	46.5	49.4	52.3	55.2	58.2
40	11.5	15.3	19.2	22.9	26.9	30.7	34.6	38.4	42.3	46.1	50.0	53.9	57.8	61.7	65.6	69.4	73.3	77.3
50	14.4	19.1	23.9	28.7	33.5	38.3	43.2	47.9	52.7	57.6	62.4	67.2	72.1	77.0	81.8	86.6	91.5	96.3
1° 00	17.2	22.9	28.7	34.4	40.2	46.0	51.7	57.5	63.3	69.0	74.8	80.6	86.4	92.3	98.0	104	110	115
1 10	20.1	26.7	36.5	40.1	46.9	53.6	60.3	67.0	73.8	80.5	87.2	93.9	100.7	107.5	114.3	121	128	134
1 20	23.0	30.5	38.3	45.8	53.6	61.2	69.0	76.6	84.2	91.9	99.6	107.3	115.1	123	131	1[illegible]	146	154
1 30	25.8	34.4	43.0	51.6	60.3	69.0	77.7	86.1	94.7	103.4	112.0	120.7	130	138	147	155	164	173
1 40	28.7	38.2	47.8	57.3	66.9	76.6	86.3	95.6	105.2	115	124	134	144	153	163	173	182	192
1 50	31.6	42.0	52.6	63.0	73.6	84.2	94.9	105.2	115.7	126	137	147	158	169	179	190	200	211
2 00	34.4	45.8	57.4	68.9	80	92	103	115	126	138	149	161	172	184	195	207	218	230
2 30	43.0	57.3	71.7	86.0	100	115	129	144	158	172	186	201	215	230	244	259	273	287
3 00	51.6	68.8	86.2	103.2	120	138	155	172	190	207	224	241	259	276	293	310	328	345
3 30	60.2	80.4	100.5	120.5	141	161	181	201	221	241	261	281	302	322	342	362	382	402
4 00	68.9	91.8	114.8	137.7	161	184	207	230	253	276	299	322	345	368	391	414	437	460

CHAIN.

As the circumstances under which the use of the chain is necessary are mentioned elsewhere, it is only requisite to give a short description of the one employed in the Coast Survey. It is twenty meters long, and consists of that number of pieces of stout iron or steel wire, exactly one meter in length, each end of which is bent into an eye, and connected by a ring with the eye of the following link. For convenience of carriage these links are subdivided in some chains; but the advantage resulting from this is questionable, as the rupture almost invariably occurs at the joints, and multiplying them increases the liability to breakage; besides, the "kinking," or tendency to overlap or double, is also increased in proportion to the number of joints. On the other hand, it may be said that the bending of the links is decreased in proportion to their shortness.

At each extremity of the chain is a large ring, which slips over a staff held in the hand of the chainman, and rests upon a projecting rim of the pointed iron shoe at its base. The centers of these rings at the ends of the extended chain are the extremities of a line of twenty meters, and the points at which the pins should be inserted during the chaining.

As the strain constantly exerted upon the chain to straighten it must finally lengthen it by the "giving" of the rings, or as it may at times be shortened when the links are bent by being drawn over fences, rocks, or other unyielding obstructions, it is well to test its length occasionally. Important errors have arisen where dependence has been placed entirely upon long chained distances from a neglect of this source of error.

Adjusting screws are attached to the terminal rings of some of the chains, by which any error of length can be corrected.

Each chain is accompanied by the usual number of pins, and a spring wire triangle for carrying them. The pins are also made of stout wire, about 18 inches long, pointed at one end, and bent into a ring at the other. It is well to attach white cotton or red flannel rags to the ring of each pin, that the rear chainman may distinguish it more readily in high grass, marsh, bushes, &c., and also at the ends of the five, ten, and fifteen meter links of the chain, for facility in counting. These rags should be renewed when they become soiled.

Care should be exercised in the selection of intelligent chainmen, since it is often upon the precision of their work that the correctness of the survey in a great measure depends, and it is not always convenient or practicable for an assistant to accompany them. The better man of the two should be placed at the rear end of the chain, as he is the more responsible, and the forward man should implicitly obey his instructions. In all important places, however, such as closely settled districts, villages, and towns, in the measurement of base lines or distances upon which anything of special importance may depend, an aid should go with the chain and check the records of the men.

When chaining is done in connection with the plane-table and the station is reached, the chain should be drawn sufficiently beyond and clear of the table to be out of the way, so that if more chaining is contemplated it does not have to pass the legs of the instrument.

With the class of young men usually employed as chainmen in the Coast Survey parties, it has been found safe and serviceable, after a little experience, in the absence of an aid, to have the rear man keep a chain-book, in which he notes all the crossings of high and low water, intersections of brooks, fences, roads, &c., with rectangular offsets to all reasonably accessible points on either hand, including bottom and tops of slopes, the record of tallies, and such other matters as may be needed, and may serve to assure accuracy in case of unfavorable intersections.

It is the habit with some, in obtaining short distances to comparatively unimportant objects, to resort to pacing, and practice enables one to ascertain thus the distances sufficiently close for plotting on a $\frac{1}{10000}$ scale, but these distances should never be great, and the topographer should be well assured of the accuracy of the pace. The use of the telemeter, however, may take the place of this method.

All distances to objects on either side of the chain line should be taken by offsets at right angles to it, and the book of the aid should, as far as possible, be so held that the line drawn as his guide in sketching should be in the same direction as the chained line itself, the better to enable him to draw his objects in their true relative positions.

Where great accuracy is necessary, the length of a gradual slope may be measured and the angle of inclination taken with the vertical arc, with repetitions, and the measurement made reduced to the horizon by means of the following formula:

Let y be the length of the line measured upon the slope, δ its angle of inclination, and its length reduced to the horizon; then—

$$x = y \cos \delta.$$

The excess of y over x may be computed by the formula, $y - x = 2y \sin^2 \frac{\delta}{2}$.

On very large scales, when parts of a meter are perceptible, Payne's tape, consisting of a narrow steel ribbon, which can be marked for minute distances, has been employed with advantage. It is convenient, also, for rapid reconnaissance in military surveys, and has been used under fire almost at double-quick. Under these circumstances three men were employed, the usual back and forward chainmen, with another to stand by the pins when stuck until about half the chain had passed, then by pointing to indicate the position of the pin to the back chainman, and run forward in time to find, without difficulty, the forward pin, and also to change the pins at each tally. This tape will not stand as rough handling as the chain, and cannot be repaired in the field when broken, while the latter can lose one or more of its links and still be of service. The easy obliteration by attrition of the marks measured upon it has also been found a source of difficulty. All things considered, however, it is a very useful, compact, and handy instrument.

TELEMETER.

In consequence of some of the disadvantages resulting from the employment of the chain, among which are the necessity of frequent dependence for correct distances upon the chainmen, the number of persons required, the time consumed in its management, and the impediments to its use found in the features of some sections of country, another instrument, styled the telemeter, has been advantageously introduced in the topographical work of the Coast Survey.

It appears that instruments of this class were at first generally regarded by scientific men as merely ingenious inventions, and not as valuable in most respects as the ordinary method of chaining, the filling in of details forming a principal exception. From the experience of its use by the officers of the Coast Survey, however, it has been satisfactorily ascertained that the rapidity with which the details of a survey can be determined and sketched by its use, the smaller number of men necessary to be employed, the advantage that the topographer observes the distance without depending upon the correctness of others, and the facility with which it may be used in places where the use of the chain is impracticable, or at best difficult, render the telemeter a very important acquisition. It is not presumed that it will ever entirely supersede the chain as a measuring instrument, but it is undoubtedly a facile and useful substitute under certain conditions.

The telemeter, as used in the Coast Survey, is simply a scale of equal parts, painted upon a wooden rod about 10 feet long, 5 inches wide, and 1¼ inch thick, so graduated that the number of divisions upon it, as seen between the upper and lower horizontal wires of the telescope, is equal to the number of units in the distance between the observer's eye and the rod held at right angles to the line of sight.

In all cases the telemeter should be graduated experimentally for the particular instrument and eye of the observer who has it in use.

The horizontal wires in the diaphragm of the telescope should be accurately adjusted, and the divisions of the telemeter made to correspond in length with the distance included between the upper and lower wires of the telescope at a carefully measured distance, and then divided into as many equal parts as there are units in the distance measured.

For convenience of transportation it can be hinged in the middle, and secured on the side when in use by a sliding bolt; and as it is necessary, when observed upon, that it should be held accurately at right angles to the line of sight, a small brass moveable bar, with sights or a groove upon its upper edge, should be fixed upon the side of the rod at a convenient height for the eye of the rodman, and which, when in position, will be perpendicular to the plane of the telemeter and directly in the line of sight of the telescope.

The correctness of the telemeter depends upon the closeness of the reading, and the accuracy with which the rod is held perpendicularly to the line of sight.

With ordinary care an error of reading should not occur even at the greatest distance denoted on the rod. With the observations carefully made, and the reading of the rod reduced to a horizontal plane, the greatest distance given by it—as usually divided—can be relied on as practically correct. There is no sensible error at any distance greater than 20 meters and less than 260, and, generally speaking, the telescopes of the Coast Survey alidades have not sufficient reading power beyond 400 meters, but it will generally be safe to rely upon it for any distance from 15 to 300 metres, beyond which it cannot be read with accuracy for use in constructing a map on a scale of $\frac{1}{10000}$.

The telemeter has been recommended for use in a great variety of cases where it becomes necessary to determine distances, in such close filling in as the corners of streets, wharves, &c., determination of all classes of detail, in traverse, shore line, and even the establishment of positions; but in the latter it is safe only to depend upon good intersections. It has been employed, however, in all manner of detail, and is preferred by some to the chain in all cases save in compactly built streets and on long lines, where the distances are so great that the telescope will not admit of the accurate reading of the rod; it is maintained by some that where only a single point is to be seen positions can be readily and accurately determined.

For the reduction of the hypothenuse to the base, the following table is given:

Table for reduction of hypothenuse to base.

Angle.	Hypothenuse.				
	100 meters.	200 meters.	300 meters.	400 meters.	500 meters.
5°	99.62	199.24	298.86	398.48	498.10
10°	98.48	196.96	295.44	393.92	492.40
15°	96.59	193.19	289.78	386.37	482.96
20°	93.97	187.94	281.91	375.88	469.85
25°	90.63	181.26	271.89	362.52	453.15
30°	86.60	173.21	259.81	346.41	433.01
35°	81.92	163.83	245.75	327.66	409.58
40°	76.60	153.21	229.81	306.42	383.02
45°	70.71	141.42	212.13	282.84	353.55

RECONNAISSANCE.

The term reconnaissance as applied to topography is a somewhat indefinite one, or rather it might be said it is very comprehensive. When there is any deviation from the closest attainable accuracy in a finished plane-table sheet, it becomes, strictly speaking, a reconnaissance map; and the rudest sketch of a country in which the features are delineated in rough approximation, which for certain temporary purposes is all that is needed, is likewise so called; so that in executing this kind of work with the plane-table there is much left to the judgment of the topographer. The amount of accuracy and closeness of detail required depends solely upon the object for which the survey is undertaken, and the time and expense allotted to its execution. It is always best, however, to strive for the greatest precision which the circumstances will allow, particularly as the sheet may at some future time become available for more important uses than that originally intended.

To the practical surveyor it is unnecessary to give any rules for his guidance, as his knowledge of the plane-table and of the requirements of the special work which he is called upon to perform will enable him to execute it promptly and satisfactorily. To the beginner, however, a few words on this subject, together with a statement of some of the results of the work accomplished by the Coast Survey officers, may not come amiss.

The recent war has shown in a forcible manner how little accurate information there was with respect to the topography of the interior of many of our middle and southern States, and the demand for an increased number of topographers in the army was supplied, in answer to the calls of the War Department and various generals in the field, from the Coast Survey; and in almost every field of operations from the commencement of the war to its close, the plane-table was used.

Until this time very little use was made in this country of the plane-table as a reconnoitering instrument, and it is the testimony of all the officers of these parties, as the result of their labors, that for rapidity and accuracy in the execution of military reconnaissance it is more effective than any other instrument at present used.

The usual system adopted, in default of triangulation, was the measurement of a base with an ordinary chain and triangulating with the plane-table.

In detailed surveys for the army, where a topographer averages about a square mile a day, a chained base of from one-half to three-quarters of a mile for the survey of an area of twenty-five square miles is found sufficient.

At Chattanooga, from two different bases of about half a mile each, plotted on separate sheets, and measured once carefully with the common twenty-meters chain, the same chain being used for both measurements, after considerable intermediate plane-table triangulation carried on by two officers, two objects were determined two and a half miles apart, common to both sheets, which were on a scale of $\frac{1}{10000}$, and the discrepancy was but about fifteen meters. Many other points of junction indicated this to be the maximum error. In this case the leaves were mostly off the trees, and the hills afforded good points. The sheets covered about twenty square miles each. At Nashville there was a discrepancy of about ten meters in two miles. This would be too much error for finished work, but it is very accurate under the circumstances.

At other times, when the character of the country or the pressure of time did not admit of the measurement of a preliminary base and topographical triangulation, the work was commenced by starting from a single point, and prosecuted by linear measurement with the chain, intersections from the ends of the chained lines being taken to determine objects, which, as the work progressed, could also be used as checks upon the chaining. Where circumstances permitted, an occasional return with the chain to a back point, either to close a series of lines upon it or to start afresh, was resorted to. This work was generally carried on over roads, and the interior filled in by intersections and sketching, as far as practicable. Some of the tests in this latter work, where the operations of two officers joined, were remarkably satisfactory.

It is estimated that with the usual number of hands and a good sketcher for aid, in a country of average variety of detail, in which all the houses, prominent barns and out-buildings, streams, roads, general outline of woods, and approximate twenty-feet contours are to be shown, on a scale of $\frac{1}{10000}$, an area of between two and three square miles can be filled in daily, with not only sufficient accuracy for military purposes, but so that a trained eye would not discover any marked discrepancy between the representation and reality. This rapidity of work, however, could not be expected in or near towns or populous districts. It is doubtful if the average work would reach more than one-half this amount.

In some thickly wooded sections, and where time is limited, it has been found advisable to run the main roads with the plane-table, and fill in with the compass, which is more rapid but less accurate than where the work is done with the plane-table alone. The usual method employed, where these instruments were combined, was as follows: Where the army was stationary, or moving leisurely, one main road was run with the plane-table, the topographer being accompanied by assistants well practiced in the use of the compass. Upon arriving at any important road or water-course, an assistant was sent to the right and left, starting from a plane-table point, determined by the chaining, and running as far as was requisite and then returning to the main road again to repeat the operation, the compass notes being kept in a book prepared for the purpose. Prominent points determined by the plane-table were used as checks in the compass work. The intervening topography, where no compass or plane-table work had been done, was sketched in by the chief of the party, in which accurate pacing became of great service.

OFFICE WORK.

All the drawing of the topographical features of a survey upon the chart should be penciled in the field, while they are still under the eye. Sketching and plotting in the office from notes, unless the country be near at hand for ready reference in case of doubt or a defective sketch, is objectionable. Where this is unavoidable the sketch should be transferred to the sheet as soon as possible after having been made, while it is fresh in the mind of the person by whom it was made, and by whom also, if possible, it should be plotted. Days which, from inclemency of the weather, are unfavorable for out-of-door work, should be allotted to this purpose, and advantage should be taken of them, also, for retouching any details of the sheet which may have become indistinct, as it is very important that they should not be left indefinite or become obliterated; for when the inking is done, as it generally is, at a distance from the field of operations, the necessity for this care is obvious. Nos. 4 and 5 pencils are good for this purpose, for which very hard or very soft and black pencils are equally unsuited.

In the inking of a topographical sheet three requisites to its proper appearance when finished should be borne in mind: clearness, neatness, and uniformity.

The lines and objects should be clean and sharply defined, nothing being left obscure or doubtful; the paper should be kept unsoiled, and erasures avoided as far as possible, and the style and strength of the drawing should be the same throughout. It is an important matter that an easy and natural appearance should be given to the map, for, as before remarked, a mere rigid adherence to conventional signs is not all that is necessary; while there should be no deviation in this respect, at the same time the draughtsman should strive to *represent the country*. There is a great difference with regard to this among topographers. Two equally correct charts of the same section of ground, executed by different persons, may be inked, and while one will have a stiff and ungraceful look, the other will appear artistic and natural, giving at once the impression of a faithful representation of the country surveyed.

Office work should not be commenced until the field-work is entirely completed, as no inked or partially inked chart should ever be used in the field. Sometimes, for the special examination of an old survey, or for the insertion of some recent artificial or natural changes, this becomes necessary. There is always a risk of injuring an inked map by exposure to the weather or by using it upon a plane-table.

The inking should commence with the shore-lines, high and low water. The high-water, or shore-line proper, should, in all cases, be full and black, the heaviest lines on the sheet, and in this, as in all the rest of the ink work, the lines of the survey should be strictly adhered to, where they are distinct records of feature.

The topography as drawn in the field is supposed to be correct when the chart is finished, and no office amendments or changes are admissible. The low-water line is next drawn, not so full as the former, but clear, black, and uniform, consisting of a dotted line for sand and mud, and the conventional sign where it is formed by shells, rocks, or coral reefs.

Grass upon flats, or shoals covered at high tide, have no distinct continuous line to mark their limits, each being represented in its proper form and within its area by its conventional sign only, but the shape should be well and correctly defined. All objects between high and low water, covered at full tide, should be represented less boldly than the rest of the map, but not faintly or indefinitely.

The roads should next be inked, plainly and evenly, with parallel sides, except where the survey shows a deviation from the general width. Where a road is fenced the fence should be shown by the usual sign, and where there is no inclosure a dotted line should indicate the road-side, and then should follow the fences and houses. In drawing the latter, care must be taken that the corners and angles exhibit a sharp, clear outline, which adds much to the appearance of the map.

The general skeleton of the survey being now completed, the contours are drawn with a bold, uniform, plain red line, without break, over all the other work, following accurately the full range of level of each of the contours on the sheet.

After this comes the general filling in, by conventional signs, of sand, marsh, grass, cultivation, orchards, rocks, hachures, &c. Some practice is needed to execute the sand-work regularly and neatly. It should never be hurriedly done, though rapidity in this respect follows practice. The lines representing marsh, and the delineation of grass on the fast land, should always run in the same direction over the whole sheet, and be parallel to the top of the sheet and the title. The appended drawing (Sketch No. 32) gives roughly the conventional signs as adopted by and now used in the Coast Survey.

The most difficult part of the inking for a beginner is the lettering, which now follows, and for which samples are given, (Sketch No. 32.) It is expected that every topographer shall have learned to draw sufficiently well to ink his sheet in a clear and distinct manner, and letter it with some regard to neatness and graphic effect, as the appearance of an otherwise well-inked sheet is sometimes marred by careless or indifferent lettering.

The location of the names upon the sheet should be such as not to cover or obliterate any detail or feature of the survey, and the letters should be put in neatly and gracefully, and in point of size and form according to the specimens furnished. The title should finally follow, with such notes as may be necessary to explain any peculiarity of the sheet or survey. This title and lettering should, as far as practicable, be so placed that when the sheet is held with the top (usually the north or east end of the map) from you it can be easily read; in other words, as nearly parallel to the top or upper end of the sheet as the nature of the work will admit. All names well established and recognized in the neighborhood, both general and local, should be collected during the survey, and their correct orthography ascertained, and, in case of any doubtful or disputed orthography, a report should be given of any traditions or any authorities which may bear upon the subject. No illuminated or German text, old English, or what is known as "fancy printing," should be indulged in, but a strict adherence to simplicity should be maintained.

The minutes of the parallels of latitude and meridians of longitude should be marked in figures at the upper and right-hand ends, respectively, the degrees on the center parallel and center meridian only.

Where the buoys are determined by the topographer, and their names, colors, numbers, or kind are known, they should be lettered upon the map.

The triangulation points should also be lettered, first being surrounded by a small circle. The magnetic meridian should be drawn with a half fleur-de-lis at its head, and the true meridian, where no projection is used, with a full one.

Spotted something that isn't quite right?

Let us know and we'll fix it for future reprints.
And as a thank you for your diligence, we'll
send you a free Sothis book of your choice.

publisher@sothis.press

This page intentionally left blank

Made in United States
Orlando, FL
05 November 2026

90231073R00028